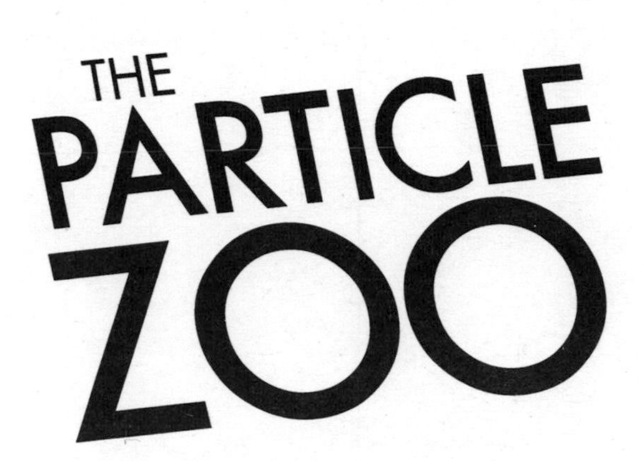
THE
PARTICLE
ZOO

The Particle Zoo

The Search for the Fundamental Nature of Reality

GAVIN HESKETH

Quercus

First published in Great Britain in 2016 by

Quercus Editions Ltd
Carmelite House
50 Victoria Embankment
London EC4Y 0DZ

An Hachette UK company

A CIP catalogue record for this book is available from the British Library

HB ISBN 978 1 78429 867 8
TPB ISBN 978 1 78429 868 5
EBOOK ISBN 978 1 78429 869 2

Diagrams by Jeff Edwards
Diagram p. 183 © CERN

10 9 8 7 6 5 4 3 2 1

Typeset by CC Book Production

Printed and bound in Great Britain by Clays Ltd, St Ives Plc

For my family

CONTENTS

AUTHOR'S NOTE AND ACKNOWLEDGEMENTS

This book is about the universe on the smallest scales. And looking on those smallest of scales, everything seems to be made of the same stuff, the most fundamental things that we know: particles. The aim of this book is to give some idea of what these particles are, how they behave, where they might all have come from, and how we know all of this. The first thing to know about particles is that they don't behave like anything else in the universe, certainly unlike anything we experience in our day-to-day lives. This is part of what makes the world of subatomic particles so interesting, but it does present some challenges when describing them.

The main challenge is that the best language used in the field of particles is mathematics. So if you really want to understand particles, you would have to learn their language – and that means maths. Having said that, I think the big ideas, the interesting concepts, can be separated from the equations and presented in everyday terms, and that is my aim with this book: to bring the subatomic world to life using analogies to more everyday things. But analogies are figurative, they have limits, and any failure of those you'll find here is mine. The mathematics is usually right!

The other choice I had to make while writing was what to include. Particle physics has been around for over a century, with many thousands of brilliant people involved in that time – and this provides a lot of material to draw on. This book contains an overview of the history, and some of the most exciting areas of research today, but it cannot be comprehensive. It is also a story of people, but here again I had to make a choice: when things are clearly history, and Nobel Prizes already acknowledge them, I have generally given the names of the people usually associated with each idea. But for current work the picture is less clear. To me it would seem to be unfair to pick out just one or two people responsible for discovering the Higgs boson, or for the theory of Super-Symmetry as we know it today. Science has always been collaborative, and nowhere is this more true than in particle physics – both on the experimental side and on the theoretical side. So rather than get bogged down in lists of names, I have decided to focus just on the big ideas in play in active areas of research.

It would also be impossible for me to thank everyone who has been involved in this book. I've really been developing these ideas throughout my research career, so everyone I've worked with, and everyone I've spoken to about the Higgs boson down the pub, has played a role. Having said that, the process of actually writing all this down is a different skill altogether, so I'd like to thank my editor, Wayne Davies, for suggesting this whole thing, and Erica, without whose love and support this almost certainly wouldn't have come together.

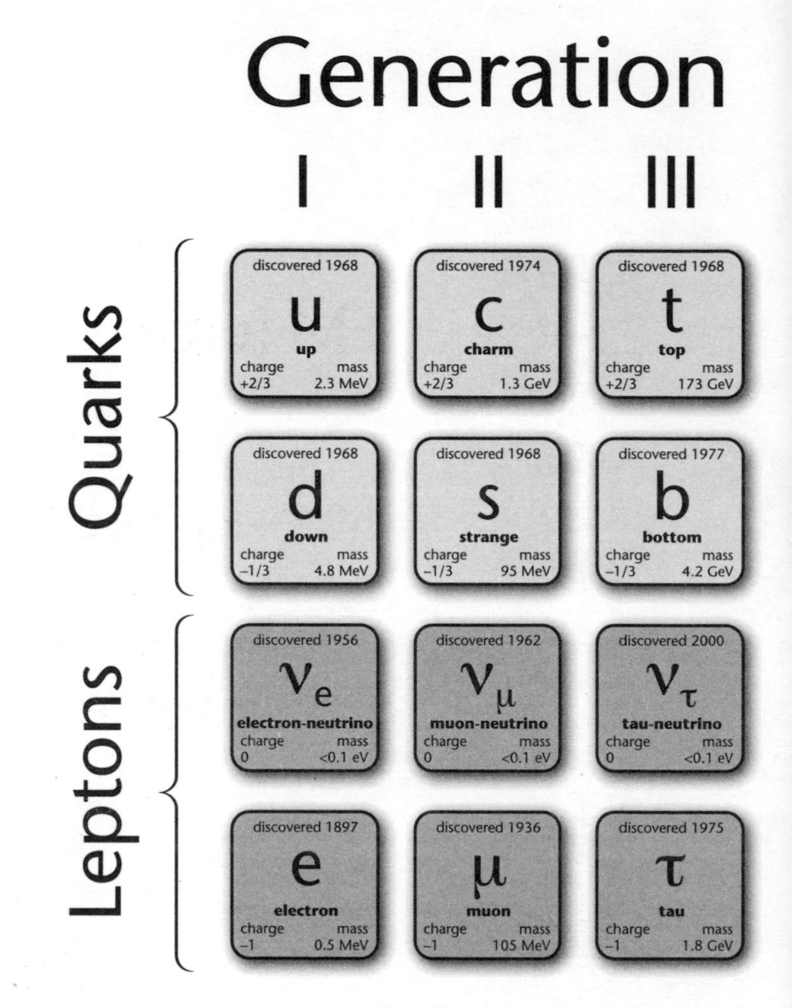

The Standard Model of particle physics

EM force

Strong force

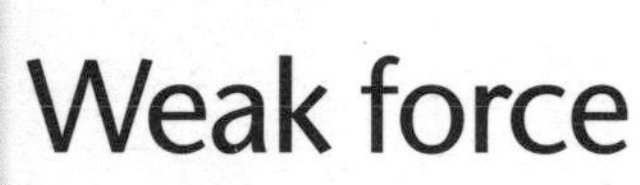

Weak force

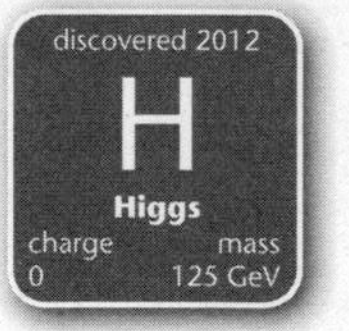

Forces (bosons)

CHAPTER 1

THE FUNDAMENTAL NATURE OF REALITY

Beauty is perhaps a strange word to use when describing science, not least because it is so hard to define. But we know it when we see it, and I was seventeen when I first saw the beauty in physics. The way certain ideas draw an unexpected connection between two seemingly different topics. How particular concepts appear over and over again. And how all of this can be expressed in a few simple yet powerful equations. When studying these ideas, it is hard not to feel that we can touch, even for just a moment, a deeper truth about the world around us.

The history of these ideas is a history of human creativity and imagination in discovering them – as well as of culture and politics, errors and luck, dead ends and breakthroughs. But the beauty of science is that it can also transcend this: a single equation really can tell us something about the universe that is as true today as it has been for billions of years, and can also tell us what will happen billions of years in the future.

But this story is not finished, and the future of science is about continuing to make sense of the world around us. Particle physics is perhaps the most extreme form of this quest, trying to understand the universe by studying the smallest things in it: the fundamental particles. The smallest pieces of matter. The basic building blocks that make up you, me, and the whole world. The same particles that form the Sun, our entire galaxy and the billions of other galaxies, and have existed for almost the entire history of the universe.

This book is about those particles. It's about what they are, how they behave, and the possibilities that exist in the world around us. But it is also about where all these particles came from, how everything began, and how it might end. These are some of the biggest questions we can ask, and the search for answers has revealed a universe that is far stranger than we could have imagined. This is the story of what we know, of what we don't, and of our quest to learn more. It is the story of quarks and leptons, of bosons and symmetries, and of the biggest experiment in history studying the smallest things that we have discovered up to this moment.

Experiments really play a crucial role in this story. Only by measuring the universe can we understand it, and the experiments I'll describe vary from the biggest, the Large Hadron Collider, a huge 27-km particle accelerator studying the most extreme conditions ever created in a laboratory, to some of the less famous, like a giant tank of dry-cleaning fluid down a mine in South Dakota which studied the heart of the Sun. Over a century of investigation has revealed an entire zoo of exotic particles, and led to the most successful scientific theory ever developed: the Standard Model. This theory describes how particles behave, how they interact, and explains everything from how atoms form to how the Sun burns, and the way these particles shape the world around us. The Standard Model has also predicted the result of every experimental test we have so far been able to devise – the real sign of scientific success.

As for me, well after getting hooked on physics I pursued a career in research at the energy frontier, the highest-energy particle accelerators in the world, finding new ways to test the Standard Model under more and more extreme conditions. Now as a lecturer in Experimental Particle Physics at University College London, I was lucky enough to be working on the ATLAS experiment at the Large Hadron Collider in 2012 when we answered the final question about the Standard Model. The missing piece that had been predicted almost forty years ago – and hunted ever since – was finally discovered: the Higgs boson. This particle started out as a mathematical trick, but eventually became the key to the whole theory – it really holds the Standard Model together, and its discovery tells us about the deeper rules governing the behaviour of the universe.

We are really in a golden age for fundamental physics: we have the greatest experiment ever built pushing back the frontiers of human knowledge. But what is most exciting is that we may be on the threshold of a completely new discovery, something much more significant than the Higgs boson, a discovery that could lead to a new revolution in our thinking about the universe and our place in it.

The atomic world

Before revolutionising physics, I need to introduce the main players in this show: the basic building blocks of the universe, and the forces that move them. This story begins with atoms.

We are taught at school that everything is made up of atoms, but this is still quite surprising when you stop to think about it. All of the dazzling array of different stuff around us, solid objects like this book, liquids like water, all the creatures on Earth and the air we breathe, everything is made of atoms; we just don't notice them because they are so small. Take any everyday object, like a piece of paper, and start slicing it up. You get smaller pieces of paper – a centimetre, a millimetre, a fraction of a millimetre in size. If we had a magic knife that could keep on slicing, we would eventually cut the paper up into some molecules, and if we sliced those apart we would be left with atoms. It's hard to really visualize how tiny an atom is, but a million of them would fit across the width of a single hair.

The idea of atoms, of basic building blocks of matter, existed in several ancient philosophical traditions, but the modern concept dates from the nineteenth century. Solid experimental evidence proved that atoms were more than an idea, from Brownian motion (the jostling of tiny pollen grains by water molecules visible with a microscope) to the behaviour of gases and the way different chemicals combine to form new compounds. In all of these experiments, atoms seem to behave like tiny solid marbles – something reflected in the word 'atom', which derives from the Greek for indivisible.There are different types of atom, and they can stick together to form molecules – like two hydrogen and one oxygen atom making H_2O, or water. Molecules can stick together to form different materials, like water molecules forming ice. Atoms and molecules can also move around and bounce off each other, and we perceive this motion as pressure and temperature: more bouncing means higher pressure, and faster bouncing means higher temperature.

If atoms behave like tiny marbles, then they must be a little solid lump of something, and it's sensible to ask what that something is – this is the search for the fundamental nature of reality after all. However, because atoms are far too small to be seen directly, it required some new experiments in the early twentieth century to learn more about them. We now know that atoms are not indivisible, but consist of a cloud of electrons orbiting a small, heavy nucleus made up of protons and neutrons – a little bit like the planets orbiting the Sun in our solar system. Building on this analogy, our solar system has 8 planets, so we can compare it to an atom with 8 electrons: an oxygen atom. If we could take one oxygen atom and enlarge it almost a million billion billion times, then the nucleus would be roughly the size of our Sun, and the rest of the atom would also look fairly similar to the solar system, just a bit bigger: the closest electrons would orbit a little way beyond Jupiter; the furthest electrons would extend far beyond Neptune. Just as the solar system is a few small planets in millions of kilometres of empty space, an atom is just a few electrons orbiting the tiny central atomic nucleus. This is one of the most surprising things about atoms: they are almost entirely empty, but can still stick together to form all of the solid objects around us. This is thanks to the way electrons in neighbouring atoms interact, and the first clue that while particles are important, it is their interactions that really matter.

Now, just as we asked what an atom is made of, we can ask what the particles inside atoms are made of, and we now know that protons and neutrons are made up of smaller particles called quarks. Two kinds of quarks, which we call the 'up' and the 'down' quarks, combine in different ways: up,up,down make a proton, and down,down,up make a neutron. Everything is made of atoms, and atoms are made of just these three particles: electrons, up quarks and down quarks.

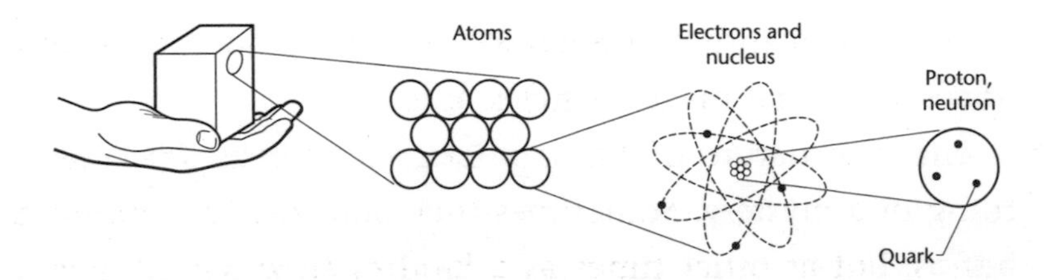

The structure of matter, from atoms to particles.

As far as we can tell, these quarks and electrons really are fundamental: they do not appear to be made of anything smaller. They may be the basic building blocks of the universe.

From atoms to particles

A fundamental particle is quite a strange thing. Take an electron: it is not a small lump of some material, because we could always ask what that material is, and an electron would not be fundamental if it is made of something else. The electron just . . . is. We don't know how big it really is, and in the mathematics of particle physics, all fundamental particles are treated as if they are infinitely small. They are a collection of different properties like mass, electric charge and so on, but with no apparent physical size.

With such strange objects, it is probably no surprise that they behave in strange ways: they live in the unusual world of quantum mechanics. This theory was developed in the 1920s, and was a true revolution in our understanding of the universe – it describes how these particles behave, and it is very different from the world as we know it. Chapter 2 explores quantum mechanics and the randomness and uncertainty that make particles odd.

The true nature of fundamental particles is still something of a mystery. Sometimes they behave like solid little lumps, but at other times as a kind of fuzzy cloud, like a tiny fly constantly buzzing around. Trapping an electron in an orbit around an atom is like trapping this fly in a bottle, and it will zip around all over the place, completely unlike the planets in the solar system, which travel in nice regular orbits. But the failure of the solar-system model of the atom was just the first of many surprises about the fundamental particles, which can travel backwards in time, bend the laws of physics, pop in and out of existence and turn into entirely different particles.

While the fundamental particles all behave strangely, they do at least have some things in common, and we use these similarities to organise the subatomic world. Electrons are a type of particle called a fermion, named after Enrico Fermi (who also has a space telescope, a particle physics laboratory, a chemical element, a few streets and several nuclear reactors named in his honour). We know of 12 different fermions, which include the up and down quarks as well as the electron, and these are the 'matter particles', the particles that make up the solid things in the universe. The basic characteristic of all fermions is that their fuzzy clouds don't overlap, but stack up instead like little plastic building blocks – the stacking of electrons in atoms gives different chemical properties to different elements, and structure to the world around us.

These fundamentally strange particles make up everything around us, but at the same time are hard to pin down. They are objects with no size, yet still take up space. We smash them together in particle accelerators to study how they behave, but the closer we look, the stranger they are. And in the rest of this book I'll explore some of that strangeness, and what it tells us about nature on the most basic level.

Forces of nature

Every particle is fascinating, but it is forces that really bring the world to life. In many ways, forces are much more familiar to us, as we experience gravity every day. We take it for granted that if we drop something, it's going to fall. Isaac Newton realised that the force that makes an apple fall from a tree is the same force that keeps the Moon going round the Earth and the planets going round the Sun, and that is the force of gravity.

The other main force that shapes our world is electromagnetism. This usually appears as separate things, electricity and magnetism, but these are really two sides of the same coin, two effects caused by the same underlying force. The electricity that powers our modern world is related to the magnetism that makes compass needles point north. In fact, without the connection between the two, we would have no mains supply: moving magnets produce electrical currents, and most of our electrical power is generated using this fact. The only difference is the way that the magnets are moved: by the wind or water, by a nuclear reactor, or by burning coal to power a steam turbine. And just as gravity holds the planets in orbit in the solar system, it is the electromagnetic force that holds electrons in orbit in the atom – without it, no matter.

Forces may be familiar, but the ways they actually work are actually quite mysterious. How does gravity pull apples out of trees? How does electromagnetism hold electrons inside atoms? In the twentieth century, our understanding of these forces was totally rewritten, providing deeper answers to this question. In 1915, Einstein's General Theory of Relativity recast gravity as a bending of space and time. In the 1920s and 30s, a new theory of electromagnetism was developed, one that works with quantum mechanics and the world of the fundamental particles. This is the nature of scientific progress: older ideas are not wrong, they are just limited, and deeper explanations supersede them. General Relativity and quantum mechanics are the two great theories in physics today, but they give two very different ideas for how forces work. Combining these ideas into a common theory for all forces remains one of the critical open problems that I'll come back to later in this book after exploring the quantum picture of the world.

Exploring the subatomic world

If the aim of particle physics were to provide a simple description of the world around us, we could almost stop here: many things can be explained with just some electrons, up quarks, down quarks, electromagnetism and gravity. But nature had many more surprises in store, and after looking inside the atom in the early twentieth century, experiments uncovered a whole subatomic world. New quarks. Forces that we don't experience directly. Particles that seem to be overweight electrons. The ghostly neutrino. It required a new way of thinking about the universe to make sense of this, and it wasn't until the 1960s and 70s that a clear picture emerged: the picture known as the Standard Model of particle physics.

The reason the Standard Model has survived to this day is that it really works, and the latest and greatest testing ground for its predictions is the Large Hadron Collider (LHC) at CERN (Conseil Européen pour la Recherche Nucléaire: the European Organization for Nuclear Research), near Geneva, Switzerland. The LHC sits in a circular tunnel around 100 metres underground, accelerating beams of subatomic particles round at 99.999999% of the speed of light, in a pipe that's both colder and emptier than outer space. These particles are smashed head-on in four different places around the circuit, where four different experiments are located: ATLAS, CMS, LHCb and ALICE. When these collisions happen, which is over 40 million times every second, they reach temperatures over a trillion degrees, and for a tiny fraction of a second they recreate the conditions of the universe less than a billionth of a second after the Big Bang.

It's almost impossible to imagine what happened 13.8 billion years ago at the Big Bang, but we know that around that time the entire universe squeezed into a tiny, rapidly expanding space. Particles were packed in so tightly that high-energy collisions filled the entire universe – so if you did ever wonder what the Big Bang was like, the LHC is the closest we can currently get. In the collisions there, we create new, exotic particles – particles that live for just a tiny fraction of a second, but that, just after the Big Bang, must have filled the whole of space. Understanding these particles tells us not just about the very early universe, but also how it evolved into the form we see today.

It's amazing to me that we can do this, and it's a huge technological achievement to make the LHC and all the experiments work. I am one of over 3,000 people from six continents working on the ATLAS (A Toroidal LHC ApparatuS) experiment, with similar numbers on CMS (Compact Muon Solenoid) – these are the two largest experiments. While any project involving this many people has strengths and weaknesses (the number of meetings comes to mind), it is a truly exciting place to be. There is a huge range of stories and motivations for doing what we do, but curiosity is one thing we all have in common. Curiosity about the world around us, and how it works. Curiosity about something that reaches far beyond our day-to-day life. But it is curiosity with a purpose: the LHC is the first laboratory in history to reach such a high energy, and the only place in the world to study what happens. We all want to be involved in discoveries, and the biggest so far came in 2012 with the Higgs boson, the final missing piece of the Standard Model picture, as we'll see later in this book.

The search

The discovery of the Higgs boson was really the end of a chapter in the story of particle physics, and we are now writing a new one. In 2015 the LHC returned after two years of upgrades, now smashing particles at even higher energies, taking us one step closer to the beginning of the universe, and to discovering previously unknown particles. But there are also many smaller experiments around the world, pushing new ideas and new levels of precision, testing things that are impossible to reach at the LHC. And all of these experiments are searching for something new. As successful as the Standard Model has been, we all want to find the point where it breaks.

And we know that this must happen, because there are problems – big questions that the Standard Model cannot answer. Telescopes measuring distant galaxies tell us that the Standard Model is somehow missing 95% of the stuff out there. We have no idea how gravity fits into our picture of the subatomic world. Or why there are 12 fermion particles in the Standard Model when it seems that only 3 are needed to make up the world around us. Current theories have taken us as far as they can, and there is no definite signpost pointing to what comes next. Later on I'll discuss some of the ideas, which include Super-Symmetry, extra dimensions, string theory and quantum gravity. One of these may combine all the different parts of the Standard Model into a

'Grand Unified Theory'. Or it may even complete the picture, giving us the instruction manual for the entire universe. A true 'Theory of Everything'.

Right now, the future is open. A Theory of Everything might be close, or nature may have many many more surprises in store for us yet – and I believe that only experiments can tell us the answer. To find the next breakthrough we must go out and discover. We are at a unique point in history, a time of pure exploration, a time for new data, new measurements and new directions. The next few years are crucial: if we do not make a breakthrough, it might be time to go back to the drawing board, and not for the first time it may require a technological leap to progress our understanding of the universe. On the other hand, we may discover a host of new particles at the LHC, a new understanding of neutrinos from the SuperNEMO experiment under the French Alps, and the true nature of dark matter at the LZ experiment deep in a mine in South Dakota, USA.

But I'm really hoping for a complete surprise, something that requires a radical change in our thinking, like the development of quantum mechanics in the early twentieth century, or the Standard Model forty years later. We are due a revolution in our understanding of the universe, and it's time to join the search for the fundamental nature of reality.

CHAPTER 2

A VISIT TO THE PARTICLE ZOO

One of the most appealing things about particle physics is that we are studying the world in its simplest form, the most basic building blocks, the things that all matter has in common. If we can understand the fundamental particles, well, it might be possible to understand everything else in the entire universe. But of course this is not a simple story, because even though these particles are all around us, they still remain something of a mystery. We don't yet know exactly what they are, and trying to find out is like trying to piece together a puzzle: we can study how they behave in different conditions, learn their characteristics, and look for clues to their true nature. And so the first thing to do is to get to know some particles a little bit, and to do that we have to spend some time with them. In this chapter I'll introduce some of the things we do know, the strange things which are true for all of the fundamental particles that make up the universe, and which are described by quantum mechanics.

The 1920s saw a revolution in our understanding of the universe, a fascinating time in the history of science as a brilliant young generation from across Europe invented the new theory of quantum mechanics. The names are synonymous with the ideas they developed: Niels Bohr (the Bohr model of the atom), Werner Heisenberg (Heisenberg's uncertainty principle), Wolfgang Pauli (the Pauli exclusion principle), Erwin Schrödinger (the Schrödinger equation), Enrico Fermi (fermions), Louis de Broglie (de Broglie waves) – and Paul Dirac, a reclusive Englishman who, in 1928, discovered an equation. It contains just 6 symbols, and describes the behaviour of an electron. But not just an electron: the Dirac Equation describes all 'fermions' – these are the matter particles, the electron and the quarks that make up atoms. And here it is:

$$i\gamma.\partial\psi = m\psi$$

It states that the mass (m) of the particle (ψ) affects how it moves (the $i\gamma.\partial$ part). I find it amazing that something so elegant and simple can describe the behaviour of every particle we find in an atom, every fermion in the universe, and every fermion that has ever existed, or ever will. This is powerful stuff. The language of Nature on the smallest scales, or at least the best language we have been able to learn so far, and the equation appears on the memorial to Dirac in Westminster Abbey, near the graves of Isaac Newton and Charles Darwin.

But of course, appearances can be deceptive – the Dirac Equation can become very complicated very quickly, and produces some surprising predictions about how particles behave. And when I imagine a particle I don't imagine an equation, I use a mix of pictures and some intuition for what that particle might be doing, how it might behave. So in this chapter I'll try to explain what a particle physicist thinks about when they think about particles, and by far the most common answer is Feynman Diagrams, so these will be our guide through the quantum world.

Cartoon physics

You can't get far in particle physics without hearing about Richard Feynman. Feynman's mathematical intuition made him one of the greatest physicists in the twentieth century, but he is also known for his science communication, bongo playing, safe-cracking and other antics. His contemporary Freeman Dyson called him 'half genius, half buffoon' after first meeting him, but later revised that to 'all genius, all buffoon'. But unlike most other brilliant physicists from the middle of the twentieth century, Feynman's name still crops up regularly in scientific meetings at the Large Hadron Collider because he was instrumental in developing the theory of particles that we use to this day. There were actually three versions of this theory, developed largely independently in the 1940s by Richard Feynman, Julian Schwinger and Shin-Itiro Tomonaga, who shared the Nobel Prize in 1965 for this work. It took a fourth person, Freeman Dyson, to realise that all three were really equivalent, three closely related descriptions of the same underlying reality. We use Feynman's method today because it is simpler and more intuitive – and comes with pictures. These Feynman Diagrams follow a simple set of rules, and give us a fairly realistic picture of how a particle moves and interacts without having to carry out complex calculations. Almost every presentation of results from the LHC will feature Feynman Diagrams, and it is not much of an exaggeration to say that all particle physicists think in the language of these pictures at least some of the time.

In this chapter I'll show how Feynman Diagrams are used to study a simple question: What happens when two electrons collide? This might not sound as interesting as making Higgs bosons at the LHC, but it will tell us a lot about how that is possible. The LHC collides protons at almost the speed of light, completely destroying them in the process. These collisions are messy, as we'll see later in the book, but we can start to build a picture of what happens by looking at the simplest type of collision: two electrons meeting head on. This situation amounts to the Rosetta Stone of quantum mechanics: it is relatively easy to study experimentally, and the findings have opened up the mysteries of quantum theory, testing ideas and predictions as the theory was developed. It also provides the basic template for the way we understand almost all particles and interactions, which we'll meet throughout the rest of this book.

First things first: we need an idea of what an electron is. In chapter 1, I mentioned that particles can do some strange things, but for now let's go back to basics, and the simplest picture for any particle is just something like a tiny marble. We already know how things like marbles move, by rolling along in a straight line, so this is how we'll start. The first Feynman Diagram represents an electron moving from A to B, and it's just a straight line between two points. In these diagrams, time flows from left to right, while the vertical direction corresponds to different positions. So far so good (a).

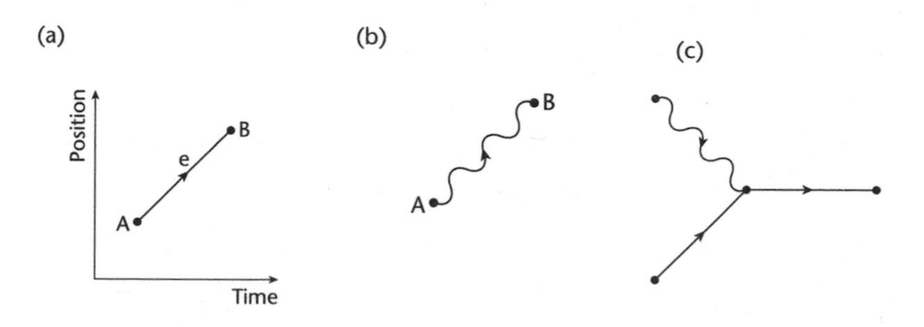

(a) The Feynman Diagram for an electron moving from A to B.
(b) The Feynman Diagram for a photon moving from A to B.
(c) The Feynman Diagram for an electron absorbing a photon, and changing direction.

Electrons don't just move, they also feel the fundamental forces of nature, and the one covered in this chapter is electromagnetism, a force which is felt by everything that carries electric charge. There are two types of charge, positive and negative. If things carry the same charge they repel; if they carry opposite charges, they attract. When two electrons collide, the electromagnetic force will make them repel each other because they are both negatively charged.

To see how this works, we need the quantum version of the electromagnetic force, known as quantum electrodynamics (QED). In QED the electromagnetic force is transmitted between the two electrons via 'messenger particles' known as bosons, and the boson associated with the electromagnetic force is actually the particle of light, the photon. In the language of Feynman Diagrams, a photon moving from A to B is also represented by a line, and to distinguish it from an electron it is shown as a wavy line and represented by the symbol γ, as shown in the middle diagram (b). (Bear in mind that this is not the same γ that appears in the Dirac Equation; Greek letters are used so often in physics that inevitably there is some recycling!)

For an example of how photons can transmit a force, imagine two people standing on skateboards. If one of them throws a ball forward, it will make the thrower roll backwards; this is the conservation of momentum. If the other person catches the ball, they will also roll away: this time in the direction the thrown ball was travelling. So, by throwing a ball between them, our two volunteers end up moving away from each other. This is what happens when our two electrons collide: they can exchange a photon, and move away from each other.

It is also possible for two particles to move towards each other by exchanging photons. If skateboarder 1 now throws a boomerang in the opposite direction, they will now roll towards skateboarder 2. The boomerang will loop around behind skateboarder 2, so when they catch it they will now also roll towards skateboarder 1. So by throwing a ball, our skateboarders can repel each other, and by throwing boomerangs they can attract. In QED, photons play the role of the boomerangs pushing opposite-charge particles together, and of the balls pushing same-charge particles apart.

We need to add one more component to our Feynman Diagrams to complete this picture: electrons absorbing or emitting photons are represented by a 'vertex', at which an electron line and a photon line combine, as shown in the third diagram, (c).

We now have everything needed to explain what happens when two electrons collide – they exchange a photon – and we can draw this as a Feynman Diagram. Reading from left to right, we can see what is happening in sequence: straight lines show electrons moving towards each other, photon being emitted, photon moving, photon being absorbed, electrons recoiling.

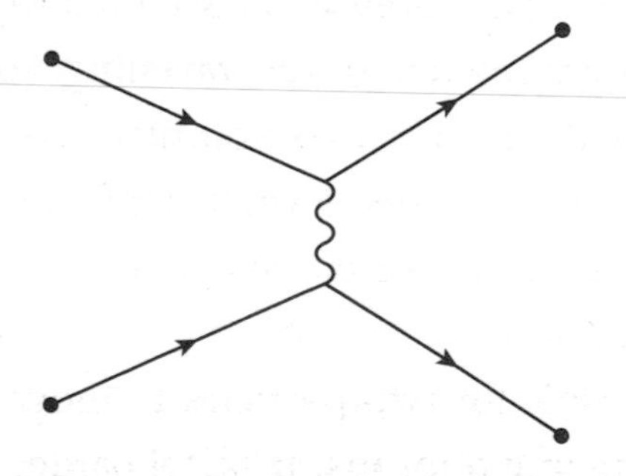

Two electrons recoiling by exchanging a photon.

Each of the component parts of this diagram represents a real piece of mathematics, and by joining them together we can build the full equation to describe ths collision. Performing that calculation is the tricky part, but drawing the Feynman Diagram that tells us what calculation to perform is easy.

This is an example with electrons, but any charged particle can be represented in the same way: by a straight line in a Feynman Diagram. With just these three simple components – particle moving, photon moving, vertex – it is possible to build the full theory of QED. They can be combined in any way, and from them it is possible to draw a picture of every electromagnetic interaction, and all of these interactions involve particles exchanging photons. Photons literally hold the world together: electromagnetic attraction keeps electrons orbiting protons in the atom, and even though atoms are almost entirely empty space, the electromagnetic repulsion between electrons in neighbouring atoms prevents those atoms from simply moving through each other. It's hard to imagine what the world would be like without photons – if I try to clap my hands, they would fly through each other without making a sound! But then without photons, there wouldn't be any atoms in the first place; the world as we know it simply wouldn't exist.

Most electromagnetic interactions are actually very complicated, involving countless photons: clapping my hands involves trillions and trillions of electrons repelling each other, and drawing a Feynman Diagram for that would take quite a long time. So quantum electrodynamics, QED, the most fundamental form of the theory, is only used in relatively simple cases, like two electrons colliding. For more complex situations, approximations like electric fields are used to calculate things, but ultimately it all comes down to QED. This theory explains how light reflects off mirrors, how electricity is generated and supplied to our homes, and how all our modern electronics work. The predictions of QED have been tested to an accuracy of around one part in a trillion, making it arguably the most thoroughly tested theory in the history of science. And it has always been found to be correct: one of science's most powerful theories really can be created from cartoons!

Playing dice

If you've ever gone bowling, you'll know there are different types of collision: a glancing blow knocking over one pin, or a solid hit for a strike. So when we collide two electrons, do they barely glance off each other, changing direction just a little bit, or do they bounce back from each other as if hitting a solid wall? The Feynman Diagram, being just a cartoon, is not very specific on this: the photon in the middle might be carrying just a little energy in the case of a glancing collision, or a lot of energy in the case of a head-on collision. The answer has to come from the underlying mathematics.

And this is where some of the unusual things about quantum mechanics start to appear. In reality, all we can calculate is probabilities: how likely it is that the two electrons just barely glance off each other, or completely change directions after a 'hard' collision, or anything in between. We cannot say for certain which one will actually happen when two electrons meet. Perhaps the mathematics tells us that there is an 80% chance of a glancing collision, and 20% for a hard collision. If we go and run an experiment where we collide electrons over and over again, we will find that this is exactly what will happen: on average 2 in 10 of these collisions will be a hard collision – if only I could bowl a strike that frequently! The problem is, just like me bowling a strike, we can't say in advance when this occasional hard collision will happen. The outcome of any collision is random, but for electrons at least, it follows probabilities that we can calculate precisely.

But the randomness in particle collisions is not the same as the randomness in my bowling. If I could measure the trajectory of the ball down the lane, it would be possible to calculate quite precisely whether it is going to be a strike or not. The randomness in particle interactions is different: even if we were able to set up an experiment where we knew everything about the electrons, the outcome would still be random. It is as if the ball only decides if it will hit a strike once it reaches the pins. Until that point, there is absolutely no way to know, no matter how precisely we measure the ball moving down the lane. This randomness lies at the heart of all quantum mechanics: for example, there is no way to know in advance exactly what will happen in each high-energy collision at the Large Hadron Collider. We can calculate incredibly precisely what is possible and how likely these different possibilities are, but then we just have to wait and see which one takes place in each collision. This turns particle physics into a numbers game. If we want to study something like the Higgs boson at the LHC, there is no way to set up a collision that is guaranteed to make one. Instead we have to produce many many collisions, and wait for the few random ones when a Higgs boson is made. And make sure we don't miss them amongst the billions of others.

This randomness troubled a lot of people early in the development of quantum theory, as it seemed to be a huge step backwards in our understanding of the universe. It is telling us that we cannot know exactly what is going to happen; not because we cannot solve the equations precisely, or because we don't fully understand the theory, but because the universe seems to be randomly choosing from a list of its own from the possible outcomes for each experiment. How is it doing this? This prompted Einstein's famous remark 'God does not play dice' – surely the universe is not really just deciding randomly what to do all the time, as if based on the roll of some dice? Niels Bohr's less famous retort was: 'Stop telling God what to do!' Quantum mechanics is unexpected, and in many ways unsatisfying, but it is a realistic description of nature. We can't tell nature how to behave, only learn how it does behave. And it seems to behave randomly.

One way or another

To get back to the example of two electrons colliding, there is a certain amount of randomness in what happens when they recoil by exchanging one photon. Maybe it is a glancing collision, or a hard collision. But there is also another possibility. Using the rules of QED, we can draw another Feynman Diagram where the two electrons interact not by exchanging one photon, but by exchanging two.

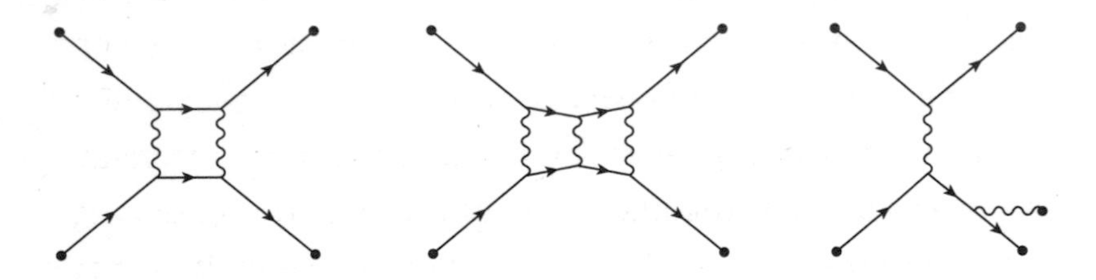

Other things that can happen when two electrons collide.

Taking a very simple approach, exchanging two photons is likely to lead to a stronger interaction. Returning to the analogy of people on skateboards, if they 'exchange' one ball, they start to roll away from each other. Exchange another, and they will roll even faster. The more photons get exchanged, the more strongly two electrons will push each other apart. There is also another possibility: the electrons may interact by exchanging one photon, but then radiate another photon later on – this photon can then travel out, like light leaving a light bulb. In this case, the energy is now shared between two electrons and one photon, so the electrons will be moving a little more slowly.

There is no way to know in advance whether the electrons will interact by exchanging one photon, or two, or even more; or if they will produce extra photons before or after the collision. All we can do is calculate how likely each of these outcomes is – and the probabilities are not equal. There is a cost to throwing a ball, or exchanging or producing a photon, and this is determined by a quantity known as the 'coupling constant'.

The coupling constant tells us how likely a charged particle is to 'couple', or interact with a photon, and it is given the symbol α. This α appears at every vertex in a Feynman Diagram, every time an electron line and a photon line meet, and it tells us what the likelihood of the electron producing this extra photon is – a bit like saying how likely it is that a coin will land on heads (which is ½). The value of α is not something we can calculate, we just have to go and measure it, and it turns out to be almost exactly 1/137. In other words, while there is a ½ chance of a coin landing on heads, there is roughly a 1/137 chance that an electron will emit one photon. The chance of two heads in a row is ¼ (which is ½ x ½), and for two photons it is roughly 1/19,000 (1/137 x 1/137).

The value of α is closely related to the strength of the electromagnetic force: if α were larger, there would be a higher probability that an electron will emit photons, and more photons would mean a stronger force pushing the two electrons apart. But it turns out that α is quite small, extra photons are quite rare, and this makes QED much easier to work with. The simplest possible Feynman Diagram for each process – known as the 'leading order' or LO part – gets us pretty close to the right answer. We can find a more precise answer by also calculating all the possible Diagrams with one extra photon – called the 'next-to-leading order' or NLO part. Very complex interactions involving many photons are extremely unlikely, so they can usually be neglected – just as we don't usually consider the unlikely result that a coin will land on its edge instead of heads or tails. But including these rare possibilities does give very precise answers: the magnetic properties of an electron have been calculated to NNNNNLO (yes, that's next-to-next-to-next-to-next-to-next-to-leading order) involving over 12,000 possible Feynman Diagrams and over 12,000 separate calculations. The result is accurate to better than one part per trillion: equivalent to calculating the distance between the Earth and the Moon to within the width of a human hair.

But there is a question here – why does α have the value 1/137 and not ½, or 1/150, or any other number? It seems as though there must be a reason for this, and some well-respected names even dabbled in numerology to try to explain it. Wolfgang Pauli, Nobel Prize winner and giant of theoretical physics in the first half of the twentieth century, struck up an unusual collaboration with the psychoanalyst Carl Jung seeking a meaning to 137. Pauli remained fascinated by this number for the rest of his life, right up to his death from pancreatic cancer in a Zurich hospital in 1958 – in room 137.

This is actually the first time we meet what is known as a 'free parameter' in particle physics, and this means that we don't yet know the reason α has the value 1/137, but once we have measured it, we can plug the number into the equations and calculate accurate predictions for how particles behave. The deeper meaning of 1/137 may come once we have a deeper understanding of the universe, but for now we simply don't know. There are actually 26 of these free parameters in particle physics, 26 numbers we cannot yet explain – in other words, there are at least 26 reasons to think that our current understanding of particle physics is incomplete.

'I said to him "You're crazy." But he wasn't.'

Up to now, things might seem a bit unfamiliar, but there shouldn't be much here that is fundamentally odd. Electrons and photons seem to travel in straight lines, and charged particles interact by firing photons at each other. We can calculate to high accuracy the various types of interaction that could take place, starting by drawing some Feynman Diagrams. There is an element of randomness, because there are several ways that things can happen, such as exchanging one photon or two, and no way to know in advance exactly what will happen, only to calculate probabilities.

But it's time for the really strange things about quantum

mechanics. These Feynman Diagrams don't just represent different possible things that can happen, they represent different possible things that do happen. And they all happen at the same time. When two electrons collide, they exchange one photon. They also exchange two photons. And three. And four. And so on. To make sense of this, it is necessary to completely change the idea of what a particle is. Up to now it has been possible to think about particles as tiny marbles rolling around, emitting and absorbing photons. In reality, particles are much stranger, and most of the time we really don't know what they get up to. Because it seems that they get up to everything, all at once.

One way to think about this is that the electrons split into multiple copies of themselves. Each copy will do something slightly different: it might move in a different way, it might fire off two photons instead of one, it might recoil in a different direction. For every possible path a particle might travel from A to B, and every possible number of photons it might fire off or absorb, there is a copy. All of these copies come back together again at the end, so that when we next make a measurement of each electron, we find just one particle again. But in between measurements, when we are not looking at them, particles seem to split into many copies, and explore all possible ways to behave. In order to correctly calculate what might happen in a collision we have to add up all these copies, and to treat the electrons as if they are doing all of these things all at the same time. Only by doing this do we arrive at a realistic description of the universe. And this is not an analogy: it is a description of the real mathematics. It is known as the 'path integral' version of quantum mechanics, because that's what the mathematics does: adds up every possible path a particle might take, every possible interaction a particle might have. This was Feynman's real insight into the nature of particles, and prompted Freeman Dyson's reaction to it: 'I said to him "You're crazy." But he wasn't.'

Because this account is so unlike anything we experience in the everyday world, it is worth exploring some more. For example, when I take the journey to work, I could either travel by bus or by train. I may not know in advance which one I will take – it might depend on which one comes first. And when I arrive at work, my colleagues may not know which one I took – but that does not mean that I took both! Of course I only took one, they just don't know which one. But quantum mechanics is different. An electron me really would split into two copies and take both the bus and the train. The electron me that takes the bus would also split many many more times, as thousands of copies of that bus spread out in all directions, taking every possible route across town. All of these different routes converge on my office, and only by accounting for them all can we correctly account for how an electron behaves.

As an aside, it is also possible to formulate quantum mechanics in another way, describing particles as waves that spread out like a ripple on a pond, in a way that is similar to the many different bus routes spreading out across town.

This 'wave–particle duality' version of the theory is easier to use for some applications, like electrons in an atom. But for high-energy particle physics, Feynman's path integral approach is much simpler – at least mathematically, if not conceptually – and this version of quantum mechanics is part of what is known as 'quantum field theory'.

If the randomness of quantum mechanics was a problem for Einstein early on, this idea was even worse. Not only can we not calculate exactly what is going to happen, we also can't know exactly what did happen – because everything happened, all at once. At least, this is what the maths says. But just because it's possible that electrons take any route, and it's possible that they emit one, two or many more photons, does it mean that they really do all of these things, all at once? Particles are so very small that we can't see directly what they are doing, so perhaps the electrons actually only did one of these and, like my office colleagues not knowing if I took the bus or train, we just don't know which one. This type of explanation is known as a 'hidden-variables' theory, as it states that there is a perfectly normal explanation for what particles are doing – we just can't see it.

One of the names most strongly associated with early quantum theory, Erwin Schrödinger, was strongly against this 'everything happens, all at once' idea, which he parodied with his famous cat-in-a-box experiment. The idea is this: a cat is in a box, along with a device that contains a lump of uranium and a vial of poison. Uranium is radioactive, and if the device detects some radiation, it will release the poison and kill the cat. Now, radioactive decays happen randomly, driven by the rules of quantum mechanics. And if we take quantum mechanics seriously, everything should happen at once: the uranium emits some radiation and the cat dies; the uranium doesn't emit radiation, and the cat doesn't die. According to quantum mechanics, the cat is both alive and dead at the same time, and one of these possibilities is picked at random when we open the box. According to a 'more sensible' hidden-variables theory, the cat is either alive or dead all along, and we just don't know which until we open the box. (This is just a thought experiment of course; no cats were harmed in the testing of quantum mechanics.)

In 1964, John Bell devised an experiment that compares the results of the hidden-variables and the everything-happens-all-at-once theories (using particles, not cats). Because there can be a difference: in the example of my quantum bus journey across town, where the bus splits into multiple copies exploring every possible route, some of these routes can end up cancelling each other out. For example, if two buses meet head-on in a narrow street, they can block each other. This can't happen in the hidden-variables scenario, where there is only one bus on the road: here no blocking can happen and the outcome for the journey can be slightly different.

Bell's idea was a way to test this possible blocking, the way that simultaneous possibilities might interfere with each other that can only happen in the quantum picture and in late 2015 the results of a new experiment finally ruled out every possible loophole. And the result? Quantum theory is right. There are no hidden variables. Particles do exist in these weird quantum states where every possibility exists at the same time, sometimes cancelling each other out. In practice, these weird states generally exist for just a tiny fraction of a second, and usually over a very small distance. Much smaller than with Schrödinger's cat, which would obviously never be both alive and dead at the same time. But in collisions like those at the LHC, which are both very small and very fast, particles behave at their weirdest. For example, there are many ways two protons can collide and produce two electrons, but there is no way we can know which of these possible events happened in any given collision, because they actually all happen in every collision. Not only is the universe random, but it does everything, all at once.

Time travel and antimatter

So far, we know that electrons can repel each other, and we can calculate precisely how likely it is that they will glance off each other or have a hard collision. We can represent this using a simple Feynman Diagram where they recoil by exchanging a single photon, a more complex Diagram where they exchange two or more photons, or really any other Feynman Diagram we can draw that obeys the rules of QED. These diagrams are useful, but only a shorthand. Particles don't really roll along in straight lines, but explore every possible path from A to B. They don't exchange just one or two photons, but every possible combination. Only by adding up all these different possibilities, all the different paths and interactions, and treating them as if they all happened at the same time, only then do we get a realistic description of what happens when electrons collide.

And, believe it or not, this all follows from the simple Dirac Equation that started this chapter. Dirac also realised that this equation predicted something else: antimatter. This does follow from the maths, but again it is easiest to see in Feynman's cartoons. Taking the vertex diagram, in which on the left side we have electron and photon lines coming together, here the electron absorbs the photon, and on the right-hand side we just have the electron line. Now, we can take this diagram and rotate it, so that on the left side we just have the photon coming in, and on the right-hand side we now have two electron lines: one going into the vertex, and one coming out.

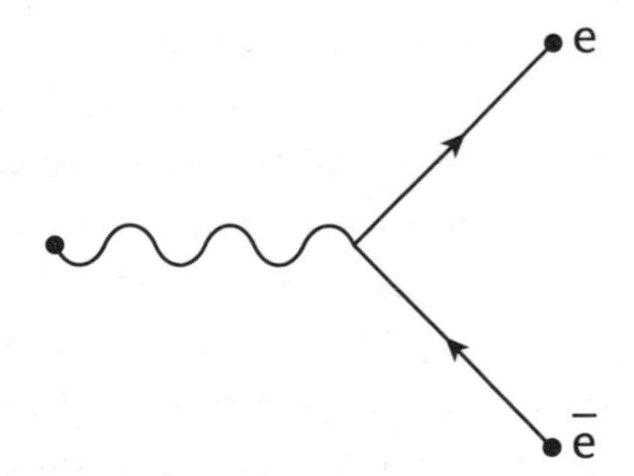

The rotated vertex diagram, showing a photon turning into an electron and a positron.

There are problems here. First of all, why is it OK to rotate this diagram? In these pictures, time moves left to right, and positions are up and down, so by rotating it we are changing what time and position (or space) mean! This is actually why the Dirac Equation is so great – it was the first equation that was compatible with Einstein's Special Theory of Relativity. Relativity tells us that all space and time are relative, and there is no absolute frame of reference: people can see things happening in different orders depending on their relative motion. It's a fascinating subject, but one for another book! Here it just tells us that there is no single way that a Feynman Diagram should be oriented, and rotating them is fine.

But there is another problem. By rotating the vertex we have moved the electron line: it was coming into the diagram from the left, but now it is coming in from the right. But travelling from right to left means it is now travelling back in time. In the Feynman diagram, this is illustrated by the direction arrow pointing backwards in time along this line, but does it physically mean anything? Well, a negatively charged particle travelling backwards in time turns out to look identical to a positively charged particle travelling forwards in time. This may sound strange, but when a negative-charged particle moves in one direction, then the place it leaves now contains less negative charge (so is more positive). Almost like an egg-timer: when we flip it over, is the sand flowing down, or the air flowing up? The line in the Feynman Diagram that we moved around now looks like a positively charged electron, called a positron. This is antimatter.

There are antimatter versions of every fermion particle: antielectrons, antiquarks, antiprotons made out of those antiquarks, and so on, and they are denoted by adding a bar over the symbol (for example $\bar{u}$, $\bar{e}$, etc), and in the Feynman Diagrams by drawing the particle moving back in time. Antimatter behaves just the same as matter, but carries the opposite charge: positrons move in the same way as electrons, and obey all the other strange rules of quantum mechanics. All this comes from just rotating the QED vertex we already know. But this new rotated vertex also tells us something else: we have a photon coming in from the left, and turning into an electron and a positron. If a photon has enough energy, it can make matter and antimatter.

To work out how much energy is needed to make antimatter we have to use what is probably the most famous equation in the world: $E=mc^2$. This comes from Einstein's Special Theory of Relativity, and states that energy (E) and mass (m) are equivalent. The other number (c) is the speed of light, and this is a big number: almost 300 million metres per second, or over a billion kilometres per hour. Square this number (multiply it by itself) and the equation becomes rather unbalanced: it means a little bit of mass can be converted into a huge amount of energy; or that it takes a huge amount of energy to make a little bit of mass.

For a photon to make an electron and a positron, we need to put twice the electron mass (the electron and positron have the same mass) into the equation, and the energy needed comes out to be a tenth of a millionth of a millionth of a joule: 10^{-13} joules (there are some very extreme numbers in this book, so I'll occasionally use this short-hand to write them; Appendix 1 has a quick guide if you are not familiar with it). For reference, there are 4,000 joules in one calorie, and about 100 calories in a banana. In other words, this is a tiny amount of energy in the grand scheme of things, but this energy is concentrated into a single photon. And just as one person having a million pounds is not the same as a million people having one pound each, concentrating even this small amount of energy into one photon makes it quite exceptional. Photons of visible light are nowhere near energetic enough to make antimatter (and this is probably a good thing). To make antimatter electrons, we have to go beyond visible light, beyond ultraviolet, beyond X-rays, up to the highest-energy photons we know: gamma rays. Hence antimatter was not discovered until 1932 because it is actually hard to make!

We can also take this rotated vertex diagram, where a photon turns into matter and antimatter, and flip it the other way around: if an electron and a positron come together, they can 'annihilate' into a gamma ray, a high-energy photon. This is the principle behind the positron emission tomography (PET) scanners used in medical imaging: some radioactive materials emit positrons, and a (completely safe) 'tracer' dose of such a material is injected into a patient. When the positrons meet electrons in the atoms in the patient's body, they annihilate to form gamma rays – actually two are produced, which fly out of the patient in opposite directions, where they can be measured. Tracking the gamma rays produces an image of where the tracer is, which tells doctors about blood flow around the body – diagnostic information that is almost impossible to obtain otherwise.

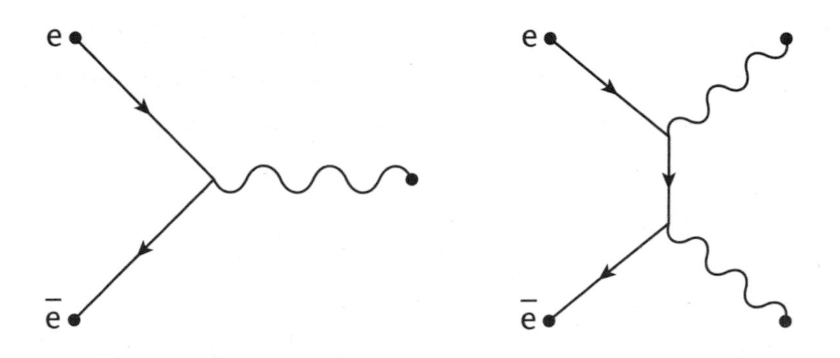

Rotating the vertex again to show an electron and positron annihilating to produce a photon. On the right, the reaction in PET scans when an electron and positron turn into two gamma rays.

This rotated vertex also allows for more complicated Feynman Diagrams: ones that contain loops. A photon may temporarily turn into a particle and antiparticle pair. Electrons may temporarily emit a photon, then absorb it again a short while later. These loops start to appear more and more as Feynman Diagrams become more complicated (the 'Next-to-Leading Order' ones and beyond). These loops tend to be very tricky to deal with, and in the 1930s almost killed quantum mechanics before it had really got going – something I'll come back to in chapter 9.

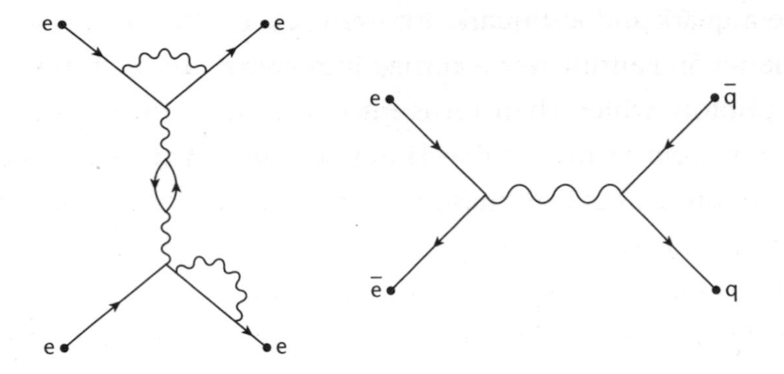

A complicated Feynman Diagram with loops. On the right, colliding electrons and positrons to make quarks and antiquarks at LEP.

Antimatter was a prediction of quantum mechanics. It just appears in the equations, and for a while it was not clear if it should be taken seriously as something real, something that actually exists in the universe, or just a strange piece of maths that can be ignored. But today antimatter is commonplace: the Large Hadron Collider accelerates particles to several million times the energy needed to make positrons, and antimatter is produced in almost every collision. Electrons and positrons are so common, and behave in such similar ways, that we tend to just call them both 'electrons'.

Before the Large Hadron Collider, CERN was home to LEP (the Large Electron–Positron Collider), a particle accelerator that actually collided electrons and positrons, and interactions there would involve matter and antimatter annihilating to make a high-energy photon, which would then turn back into matter and antimatter. But the particles at the end don't have to be the same as the particles at the beginning: it could be a quark and antiquark, for example. In these collisions, matter and antimatter is turned into energy, in the form of a photon, which then turns into different particles, a different form of matter that simply did not exist before the collision. This is like smashing two porcelain vases together and having two toasters fly out – there is no parallel to it in the world as we experience it. But all of these interactions are predicted by QED, and well understood. In fact, the only thing we consider strange about antimatter today is that there doesn't seem to be any of it left in the universe – but that's a question for later in the book!

The most precise scientific theory ever developed

In this chapter, I wanted to explain what a particle physicist thinks about when they think about particles – and the answer is usually Feynman Diagrams, the cartoon representation of the real mathematics that predicts how particles behave. There are just three basic ingredients to these diagrams: straight lines, wavy lines and a vertex (two lines coming together), and by combining these we can draw every possible way that particles can interact. Those diagrams can then be translated directly into equations that tell us the result. These are the rules of QED, the quantum form of the electromagnetic force, which is also the basic template for how we understand all the forces of nature, and all particle interactions. And generally, the 'Leading Order' diagram, the simplest one you can draw for a given process with no extra lines, is a pretty fair approximation to what is happening. But it's not the full story – there is all the weirdness too: the fact that everything happens at once, and particles can travel forwards and backwards in time. And if you accept all that, then QED can be used to make the most accurate predictions in the history of science.

I found learning the basics of QED was one of the most exciting and even beautiful things in particle physics – it is so simple, and yet so powerful. But what does all of this tell us about the particles themselves? QED describes how they behave, but it doesn't really tell us what they actually are. And it seems that particles behave in different ways: over long distances and times, they do behave like solid lumps, like tiny marbles. But over short distances and times, such as in the high-energy collisions at the Large Hadron Collider, they exhibit all of the quantum weirdness I've talked about: splitting into lots of copies, exploring all paths, and so on. So, fundamentally, what is a particle? In the end, we have only the maths to guide us, as we can never see one directly. It's like going to a restaurant and trying to work out how a meal was cooked without being allowed into the kitchen to see the recipe. We only get to taste the soup, to measure a particle at a particular moment; we don't get to see the ingredients, to see particles while they are performing all the strange quantum things.

So is the mathematics of quantum mechanics a real, factual description of what particles do when we are not measuring them? Is it the real recipe for the soup we are eating, the truth about how the universe behaves on the smallest scales? Well, we certainly can't ignore it. Quantum mechanics is the only working description of the universe that we have. It is amazingly precise, and so far every experimental test has verified its predictions. There is no going back to a simpler picture where particles are well behaved, don't split into ghostly copies of themselves, and don't travel backwards in time. And until we observe particles doing things that quantum mechanics can't predict, then there is no reason to think that it isn't the right recipe.

However we look at it, fundamental particles are fundamentally odd things. It's quite possible that quantum mechanics is really just a good working approximation to what they really are. But right now, quantum mechanics tells us what we need to know: the possible outcomes of an experiment. And it does this beautifully. Without it we would have no lasers, no LEDs and no semiconductors – and hence no computers, smartphones or any other modern electronics. No MRI (magnetic resonance imaging) scanners or PET scanners. People are now developing quantum computers, quantum cryptography, spintronics, entanglement, quantum teleportation, and there may be countless other applications yet to be invented. Quantum mechanics may have started with a few physicists trying to make sense of some unusual experimental results, but today it is essential to our lives. There is no getting away from the fact that it is weird, but most of the conceptual problems with quantum mechanics come when trying to over-interpret what the mathematics really means, what it tells us about the things we cannot measure – the events behind the curtain. But as with all theories, it is important not to stray too far from reality: quantum mechanics was invented because experiments were telling us that particles must be behaving in this way – and quantum mechanics will continue to be our best theory of particles until experiments tell us that it isn't! So that is our next stop: to explore the experiments that measured how electrons behave and opened up a whole world of exotic new particles, leading to the Standard Model, the Higgs boson, and the Large Hadron Collider.

CHAPTER 3

INTO THE HEART OF MATTER

The universe is a pretty strange and unlikely place, but you might not know just by looking at it through human eyes. It's not obvious that the world is made of atoms, or that atoms are made of electrons orbiting a nucleus. And that those electrons – the first fundamental particles we've met – sometimes behave like little marbles, sometimes like waves, and sometimes like nothing we can really imagine. They interact by firing photons at each other. They can be created and destroyed, and have an antimatter partner, the positron. How do we know all of this about electrons? The same way we know most other things: we worked out how to measure them.

Even though today we are in a golden age for particle physics, there is no escaping the fact that science can be painfully slow. It can take months or years to extract a meaningful measurement from the data at the Large Hadron

Collider, and turning an idea into a result is usually a mix of inspiration and determination. But as Isaac Newton said, we see a little further by standing on the shoulders of giants, by building a course of progress on top of all the work that has gone before. Still, while every result is scientifically interesting, some are sure to be more interesting than others. This chapter is the story of just some of the most important experiments and measurements in the history of particle physics. How a few people working with devices that fit on a table top developed into the huge research labs and giant detectors we use today. How the discoveries made in these experiments completely changed our understanding of the world. And how nature has time and again surprised us.

Physics before particles

In the late nineteenth century, it seemed as if science was almost complete. A set of ideas had built up over the previous few centuries until almost everything in the world around us had an explanation or an equation attached to it: the motions of planets, light and electromagnetic waves, and how tiny, solid atoms build the universe we knew. This set of ideas is now known as classical physics.

One of the main foundations of classical physics is Isaac Newton's *Philosophiæ Naturalis Principia Mathematica* – *Principia* for short – published in 1687 (in English, *Mathematical Principles of Natural Philosophy*). In this he sets out the Laws of Motion, which are a precise mathematical description of, well, how things move – what happens when things are pushed, pulled, or any force is applied. And he incorporated one new force into the picture: the force he called gravity. We take it for granted today, but it was truly a huge conceptual leap to relate an apple falling from a tree to the motion of the planets, which appear as just a few spots of light in the night sky. To arrive at these laws, Newton had to invent a new type of mathematics: calculus (in a pattern we shall see repeated later, calculus was simultaneously and independently invented by the German mathematician Gottfried Leibniz, and it is his notation we use today). The use of the word 'Law' suggests how astounding these ideas were at the time. To be able to distil the confusing, messy world around us into just a handful of simple equations – surely this is telling us something deep and true. A Law of Nature.

Still, Newton left some work for his successors. The Laws of Thermodynamics followed the industrial revolution in the eighteenth century, introducing the concepts of entropy and work; to put it simply, how much useful energy you can really get out of a steam engine. The field of chemistry was also making huge progress at this time. The idea that everything is made up of atoms was widely accepted, and atoms were thought to be solid little balls. Many different kinds of atom, corresponding to different chemical elements, had been discovered, and in 1869 Dmitry Mendeleev published the first generally accepted Periodic Table. This grouped the known elements by their chemical similarities, and predicted some as yet undiscovered elements to fill in the gaps. Modern versions of this table are now a standard in every science classroom.

Finally, there remained the phenomena of magnetism and of static electricity. Though they had been known for thousands of years, in the 1820s Michael Faraday carried out a series of brilliant experiments that showed how these two things were related. Based on the results, James Clerk Maxwell built his theory of electromagnetism in 1862. Why a 'theory' and not a 'law'? Just as with modern theories, Maxwell saw his equations as a model, a description of the world, rather than the true rules that make the universe tick. But what a description! His theory unified electricity and magnetism in four short equations, and predicted the existence of electromagnetic waves: ripples of electricity and magnetism that correspond to radio waves, visible light, gamma rays, and everything in between. Only when looking very closely at Maxwell's waves do we see that they are actually made up of particles, of photons.

Classical physics really reached its peak when Einstein resolved some inconsistencies between Maxwell's electromagnetism and Newton's Laws, leading in 1905 to the Special and in 1915 the General Theory of Relativity – the theory of gravity that we use to this day. But by then the quantum revolution was already starting, and everything known about the physical world was being turned on its head. Einstein himself had played a key part with the first paper about photons in 1905, though later he could never quite accept that the beautiful world of classical physics had been replaced by the randomness of quantum mechanics.

Nobody could have seen this coming. In the late nineteenth century, physics had just a few minor details left to be explained, certainly nothing to suggest a revolution was on the horizon. But scientific progress is often driven by new technology, and new experiments allowed these minor details to be studied properly. A tug on the few loose threads of classical physics undid the whole tapestry, and quantum mechanics and particle physics were developed to fill the gap. It was the late 1960s before theoretical physics truly caught up, when the Standard Model finally provided a full description of what the experiments were measuring. The Standard Model has survived largely intact to this day, as its predictions are confirmed by experiment after experiment, culminating with the discovery of the Higgs boson. This was the end of a chapter of over a century of innovation and insight, a story that starts with the particle accelerator that used to live in almost every front room.

Mysterious rays

The modern television is pretty amazing: bright colours, high resolution, and most of all, very slim and light. It was not always this way. Big, bulky and heavy, televisions all used to contain a CRT, or cathode ray tube, a sealed glass unit with most of the air pumped out, and a screen at one end. Applying an electrical voltage across the tube causes a glow to appear on the screen, and with modern (or at least mid-twentieth-century) electronics, this glow can be produced on demand and controlled to provide a moving image.

Predecessors to the CRT were being made in the nineteenth century, when they were rather small and fragile things. Building these vacuum tubes pushed the skills of glass blowers and instrument makers to the limit, but by the 1870s William Crookes had perfected a design rather like a modern fluorescent light. These Crookes tubes produced a clearly visible glow at one end when connected to an electrical power supply, and this was thought to be caused when some 'rays' flew from the negative voltage end of the tube (the cathode) and hit the other end. Whatever these rays were, they were either completely invisible or simply far too small to be seen directly. Were they one of Maxwell's electromagnetic waves? A molecule? An atom? Nobody knew, but they were the loose thread that started the revolution.

In 1895, the German physicist Wilhelm Conrad Röntgen was experimenting with a Crookes tube, and noticed something else: under certain conditions, a different kind of ray seemed to fly out of the tube and across his laboratory. Unlike the cathode ray, these new rays could also pass straight through some materials, and could even mark a photographic plate (standard equipment in a nineteenth-century laboratory). Roentgen found they could be stopped by dense materials like bone, but not by light materials like skin and muscle. Placing his wife's hand in front of a new photographic plate and turning on the equipment meant that the plate was exposed to the rays everywhere except behind the bones in his wife's hand. Upon developing this skeletal image, she pronounced 'Ich habe meinen Tod gesehen!' ('I have seen my death!') In many countries these rays are still referred to as Röntgen rays, and we now know they are high-energy photons. But to Röntgen they were a mystery, so he named them X-rays, as X is the mathematical symbol for unknown.

The following year, 1896, the French physicist Henri Becquerel discovered that certain uranium salts were spontaneously emitting some other kind of rays that could also leave images on photographic plates. Marie and Pierre Curie went on to discover and study completely new elements that also emitted rays, a phenomenon that Marie named 'radiation'. People were starting to take notice, and although rays were appearing in more and more places, there was still no clear idea of what they actually were.

Then in 1897, Joseph John (J. J.) Thomson used a Crookes tube to make a series of measurements of cathode rays. Even though most of the air was pumped out of the cathode ray tubes, there still remained plenty of atoms bouncing around inside. The puzzle was how the cathode rays were able to travel the length of the tube without being deflected. He concluded that they must be particles, and like a cyclist weaving through rush-hour traffic, they had to be small – much smaller than an atom. Thomson named these tiny particles 'corpuscles', but it was the name 'electron' that stuck. Up to that point, atoms had been thought to be solid and indivisible: the smallest building blocks of the universe. But Thomson had proved that cathode rays were something much smaller.

At the time, Thomson was a professor at the University of Cambridge, and one of the students there was a young New Zealander, Ernest Rutherford, who began to study the rays discovered by Becquerel. He realised that these rays were actually two different types of particle, which he named 'alpha' and 'beta', and that they could pass through different amounts of material before being stopped. We now know that alpha particles are fragments of an atomic nucleus, and beta particles are electrons; but at the time, nobody knew that atoms even had a nucleus: they were still thought of as completely solid, or something like a plum pudding, a solid mass with Thomson's electrons dotted throughout.

By 1909 Rutherford had won a Nobel Prize and moved to the University of Manchester, where he supervised Hans Geiger and Ernest Marsden in an experiment testing what happened when these alpha particles were fired at a gold foil. Most flew straight through the foil as expected, but moving the detector to the other side of the foil they realised that some were bouncing back. A few alpha particles were hitting something very heavy. And the thing they were hitting must be very small, considering how few alpha particles were affected. This simply shouldn't happen with the 'plum pudding' model of the atom, and Rutherford realised that the atom must look more like the solar system I described in the first chapter: a small, heavy nucleus in the centre, with electrons orbiting around it. The few alphas that hit a nucleus bounce back; most just fly straight through the clouds of electrons without being deflected at all.

Rutherford's model of the atom matched what was seen in the experiment, but could not be explained by classical physics. According to what was known at the time, these solar-system atoms simply shouldn't exist, as the electron would spiral into the nucleus in a fraction of a second and the whole thing would collapse – along with the world as we know it. But the evidence was clear, and explaining the stable electron orbits would require a complete rewriting of the physics of subatomic particles, opening the door to all of the strange behaviour we met in the previous chapter. The Geiger–Marsden experiment more than any other triggered the quantum revolution.

Clouds in a box

Perhaps the most important piece of experimental equipment in the history of particle physics is also the most unlikely. It is perhaps the dream of every university and research lab: gather together enough brilliant people and brilliant, if unexpected, things can happen. And in this case, it happened to another student in Thomson's lab, Charles Thomas Rees (C. T. R.) Wilson. As a young student Wilson planned on becoming a physician until science caught his interest – meteorology in particular. In 1893 he was spending some time at the observatory on top of Ben Nevis, where his fascination with the beauty of clouds took root. While there, I can only imagine that Wilson had some thoughts that most experimental scientists have had: wouldn't this be easier in the lab? Returning to Cambridge, he set about making a cloud machine.

Clouds, like steam, are made of water droplets: water that has condensed back from a gas into a liquid. This usually happens when evaporated water hits colder air, as when your breath fogs up on a frosty day. Clouds in the sky usually form at a certain altitude, where layers of air of different temperatures meet, with a colder layer on top of a warmer layer – moving up across this boundary is often the bumpiest part of any aeroplane flight.

To set about making a cloud he could study in the lab, Wilson wanted to evaporate a lot of moisture into the air, then cool the air fast so that the moisture would condense. He achieved this by rapidly dropping the pressure. A similar effect happens when spraying an aerosol: the contents are compressed in the can, then expand and cool rapidly when sprayed. To make this work, Wilson built a 'cloud chamber', consisting of a sealed glass vessel connected to a piston. Inside the vessel, the air was saturated with water, and moving the piston out rapidly expanded and cooled the air, causing a cloud to briefly appear in the chamber. And all small enough to fit on a table top.

But cooling alone is not enough: condensation needs a trigger. A similar effect can be seen with any fizzy drink:

carbon dioxide gas dissolved in the liquid doesn't all immediately escape, it needs a trigger to do so. Drop a grain of salt into the drink and it acts as this trigger: carbon dioxide molecules clump around the crystal, and it will leave a trail of bubbles. For clouds in the atmosphere, the triggers are typically tiny grains of dust, and water molecules clump around these grains until there are enough to form a droplet. In order for the cloud to form in Wilson's chamber, there had to be a trigger. And while experimenting, he tried purifying the air so as to remove all dust and impurities. Expecting no clouds to form under these conditions, he found a puzzle: the cloud still appeared. But what was the trigger for the water to condense?

And this is how we get from clouds to particles. Working in the same laboratory as J. J. Thomson, Wilson knew about the research into rays and radiation, and had the idea that these might be providing the seeds for the cloud. Putting the chamber near a source of X-rays confirmed it: a much thicker cloud formed. This discovery might have gone nowhere, as Wilson would spend the next decade studying real clouds. But in 1911 he improved the design to the point where it became possible to see the vapour trails of individual particles as they flew across the chamber, much like the contrails of aircraft high in the sky. At this point nobody had directly seen an atom, let alone a subatomic particle, so when Wilson published photos of the trails he was seeing in his cloud chamber, it came as a revelation.

Seeing particles

At this point, it is worth explaining what was actually going on here, because it is the same basic principle behind almost every modern particle detector. We now know what all of these rays are: X-rays are high-energy photons, alpha particles are fragments of atomic nuclei, and cathode rays and beta particles are both electrons. As any of these rays move through the air, they tend to hit some of the atoms in the air. And because atoms are mostly a large cloud of electrons around a tiny nucleus, it is usually the electrons that get hit.

In the previous chapter, I explained what happens when a photon like an X-ray meets an electron: the electron absorbs it, and receives a jolt of energy. This jolt is usually enough to 'ionise' the atom: knocking the electron out of orbit, leaving a free negatively charged electron and a positively charged atom, or ion.

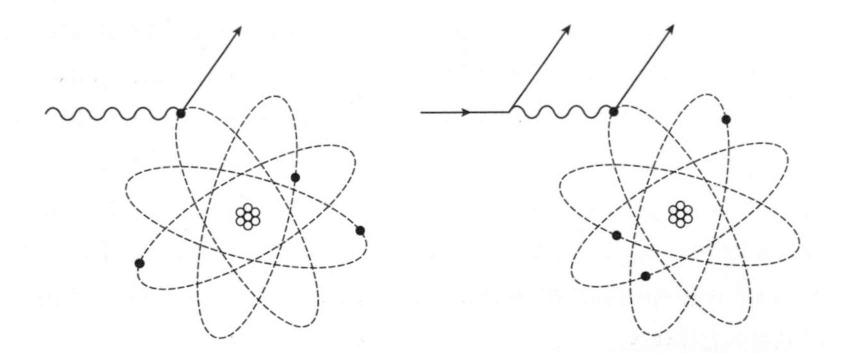

Ionisation of an atom by an X-ray on the left, and a charged particle on the right.

A similar thing happens with charged particles, such as the accelerated electrons in cathode rays. Again, we already know what happens when two electrons collide: they exchange one (or more) photons, and recoil from each other. Charged particles moving at high speed can knock many electrons out of atoms as they travel, much like a bowling ball knocking over pins as it rolls along. An X-ray photon will deliver all of its energy to a single electron, but an alpha or beta particle will leave a trail of ionised atoms, flying along until it runs out of energy.

This ionisation was acting as the trigger causing the cloud in Wilson's early chamber. Water molecules clump around the ions, and eventually form droplets. And in the 1911 refined model, he was able to control the conditions enough to identify the individual trails of droplets forming along the trail of ionisation left by charged particles.

It's possible to make a low-budget cloud chamber using an empty fish tank, with dry ice instead of a piston to provide the cooling, and this is one of my favourite particle physics demonstrations (there are plenty of guides to building one online if you would like to see some particles yourself). The droplet trails left by charged particles appear like fine lines of spider silk, floating in a cloud for a couple of seconds before disappearing. Wilson's first photographs of particle trails are now in the archive at the Royal Society, and look remarkably similar to what you can see in a dry-ice cloud chamber today.

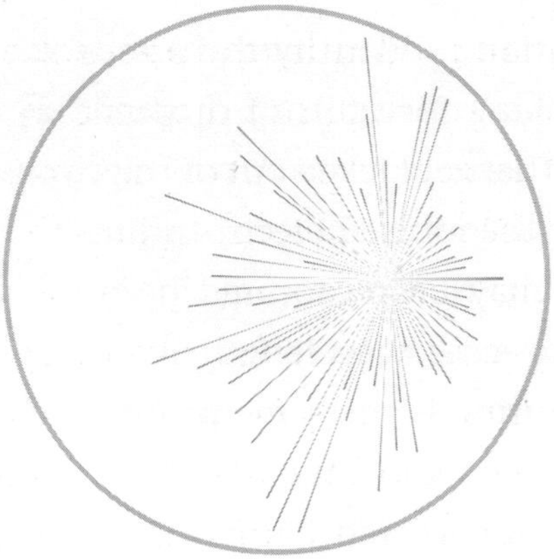

The tracks of particles flying out of a radioactive source, as they would appear in Wilson's cloud chamber.

Wilson's photos of particle tracks actually posed a challenge to early quantum theory. Heisenberg in particular had been treating particles as a kind of mathematical abstraction, something that could never be directly observed. Yet here they were, rolling along like little marbles, leaving a trail as they went. What did this mean for the quantum picture? Well, these trails are not quite what they seem. Rather than a continuous track of a particle, they are actually a series of collisions between a particle and different atoms. These collisions may happen within a fraction of a millimetre of each other, but for a particle this is a huge distance – equivalent to a golf ball travelling well over a kilometre before hitting anything. When looking at a trail several centimetres long, particles do seem like well-behaved, solid things. But that is because all of the quantum weirdness happens over very much smaller distances, in the individual collisions.

For decades the cloud chamber was the only way to see particles, and even modern particle detectors still rely on ionisation to identify the presence of a particle. But instead of looking for trails of droplets, we now look for the electrons that get kicked out of the atoms. Some detectors simply count the ionised electrons directly. Alternatively, the electrons may be recaptured by an atom, releasing a photon of a specific wavelength (colour) at the same time. This recapturing is the same process that causes the glow in the cathode ray tube, the orange in sodium street lights, and the different-coloured lights that make up the aurora near the North and South poles. Some particle detectors count these photons. Either way, the collected photons or electrons tell us where a particle was, and how much energy it had. The experiment I work on, ATLAS at the LHC, makes these measurements with over 100,000,000 individual sensors in a huge detector over four storeys high – technology that would have been unimaginable to a physicist working in 1911, even if the basic ideas are not so very different.

Rays from outer space

The experiments by Thomson, Rutherford and many others had used rays to open up the world of the atom in the early twentieth century. But it was another type of ray that would open the door to the exotic particles we study today. A ray from outer space.

Following the discoveries by Marie and Pierre Curie, it was known that there is ionising radiation all around us, and it occurs naturally in certain rocks and minerals. Get further away from the rocks, and you should see less radiation. Science is about testing your assumptions, so in 1909, the German scientist Theodor Wulf set about measuring the amount of radiation at the bottom and the top of the Eiffel Tower. He found a surprise: the amount did not fall anywhere near as much as he expected as he moved up the tower and further from the Earth. This was taken further by the Austrian scientist Victor Hess, who in 1911–12 took some radiation measurements on balloon flights over Vienna and what is now the Czech Republic. Reaching altitudes of up to 5,300 m, where the temperatures can drop to -20 °C, Hess made precise measurements with an unexpected conclusion: the higher he went, the more ionising radiation there was. To rule out the Sun as the source of this radiation, he took some flights at night, and even during a partial solar eclipse. The radiation persisted. Hess had discovered a new source of radiation coming from outer space: cosmic rays. These seemed to be something completely different from any of the rays known up to that point, which could only travel short distances. These cosmic rays were travelling many kilometres through the atmosphere, many even reaching the surface of the Earth. But as news of this discovery spread, Europe was plunging into the chaos of the First World War. Some scientists dedicated their expertise to helping the war effort at home, and many were killed in the trenches.

After the war, physics played a role in reconciliation when in 1919 the English scientist Arthur Eddington set out to test the recent theory of a German scientist, Albert Einstein. Eddington's measurements of the positions of stars during a solar eclipse confirmed that gravity can bend light, one of the strange predictions of General Relativity, launching Einstein to global celebrity. Just a few years later, the American astronomer Edwin Hubble made the measurements that proved for the first time that the universe was larger than our Milky Way – there were not just other stars out there, but other whole galaxies, and lots of them. Following the chaos of the war, the universe became a whole lot stranger, and a whole lot larger.

By the 1920s, interest in the cosmic rays discovered by Hess was also picking up, and the cloud chamber was the perfect tool with which to learn more about them: if these rays were made of charged particles, they would leave visible tracks. Wilson's design was now being used and developed at universities around the world, and one of the most important changes was the addition of large magnets. A charged particle changes direction as it flies through a magnetic field, leaving a curved track in the cloud chamber. Positively and negatively charged particles will curve in opposite directions (bending left or right, for example), and slower particles will curve more than faster particles, allowing much more information to be extracted from cloud-chamber photos. Huge powerful magnets are a key part of almost every modern particle detector for the same reasons: you can determine the electric charge and the momentum of a particle by measuring the curve in its path.

Cosmic-ray hunting was still a matter of luck though: the

cloud chamber was activated by firing off the piston to cool the air and form the cloud, then taking a photo of that cloud to be studied in detail later. If no cosmic rays happened to be passing through the chamber at that precise moment, no trails of droplets were visible. Perhaps only around one in twenty photos would yield a track, but the tracks were found: cosmic rays are made up of charged particles. In 1932, Carl Anderson, a young researcher at the California Institute of Technology (Caltech), noticed some strange cosmic-ray tracks in his cloud chamber: these seemed to be left by a particle with the same mass as the electron, but opposite charge (because the track was curving in the opposite direction in the magnetic field). This matched the strange prediction from Dirac's equation in 1928: antimatter. Anderson had discovered the antielectron, also known as the positron. Not only did cosmic rays contain charged particles, but they provided a new way to test some of the strange predictions of quantum mechanics.

Back in Cambridge, the English physicist Patrick Blackett was developing a way to make the photography of cosmic rays much more efficient, by only triggering the cloud chamber when something interesting was happening. He placed Geiger counters, simple radiation detectors, around the outside of the chamber, and used them to detect an incoming cosmic ray and fire the piston. The cloud would then form just as the cosmic rays were flying through the chamber, and now almost every photograph would capture some tracks. Blackett was able to quickly confirm Anderson's discovery of the positron, and even observed something new: 'V' tracks. These correspond to one of the vertex diagrams from the previous chapter: a gamma ray, a high-energy photon, would fly into the chamber without leaving a track, and convert into an electron and a positron, which would fly apart leaving tracks as they went, creating a 'V' shape. This further confirmed Dirac's prediction of energy (the photon) converting into matter and antimatter.

The idea of 'triggering' detectors remains essential to this day. At the Large Hadron Collider, protons smash head-on and the resulting particles fly out at close to the speed of light. Within a few nanoseconds, these particles have flown out through the detector; if it is not recording at exactly the right moment, then we miss the action. Even when everything is properly synchronised, collisions happen at such a high rate (around 40 million every second) that it is impossible to record them all, and a split-second decision has to be made whether each collision is worth saving. A whole series of triggers look for signs that interesting things might have happened: a part of the detector lighting up with lots of signals, for example. If one of these triggers fires then the data from the collision is recorded for later analysis. Get the trigger design wrong, and the interesting collisions are thrown away and lost for ever – it really is one of the most important aspects of running a particle-physics experiment.

'Who ordered that?'

In 1936, just four years after proving the existence of anti-matter, cosmic rays were to deliver a complete surprise. Back at Caltech, Carl Anderson had taken on his first graduate student, Seth Neddermeyer, and together they continued to work with cloud chambers. They noticed a track that appeared to belong to a particle much heavier than an electron, but much lighter than a proton. This was something new: it wasn't part of the atom, or any of the other types of radiation known at the time. It doesn't exist in the world around us – but it was being made by these strange cosmic rays. We now know that this particle is the muon ('mew-on', given the symbol μ), which is a heavy cousin of the electron, but at the time it was a complete surprise, and as the American Nobel Prize winner I. I. Rabi quipped, 'Who ordered that?' This was the first clue that there is more to the universe than just atoms and the particles they contain.

It was not just this muon that needed explaining, as there was still no clear indication of what was causing these cosmic rays in the first place. There were certainly some strange ideas: Robert Millikan, who won a Nobel Prize for measuring the charge of the electron, developed a semi-religious model, believing that cosmic rays were the photons emitted in the 'birth cry of atoms', as God continued to create new matter in the universe. We now know that cosmic rays are mainly protons flying through the galaxy at high speed. Many are produced by supernovae – stars that reach the end of their lives and explode in one of the most violent events in the universe, scattering light and matter in all directions. When these protons collide with the Earth's atmosphere they create a whole shower of new particles, much like a collision at the LHC, sometimes even reaching energies many thousands of times higher. For this reason, cosmic rays remain a fascinating subject of research, and will come up again in chapter 8.

But back in the 1930s, the muon was still a mystery. One theory suggested there were two types of electron: 'red' and 'blue'. One of these is the electron we know and love, the other had some rather contrived behaviour that would match this new particle. Twisting current knowledge to try to match new data is sometimes more palatable than having to deal with an entirely new reality! Then in 1935 the Japanese theorist Hideki Yukawa developed a model of the atomic nucleus that needed a new particle, called the pion ('pie-on', given the symbol π), to hold it together. The mass of the pion was predicted to be very similar to the newly discovered muon. Perhaps this was what had been discovered?

Before the mystery could be solved, the Second World War broke out. Einstein and many others fled or were driven out of Germany as the Nazis gained power. Some, like Heisenberg, decided to stay and either try to ride out the chaos or throw in their lot with the new regime. In the UK, Dirac worked on ideas for an atomic bomb, while the Manhattan Project in the USA pulled in many of the brightest minds, including Feynman, to successfully build one, unleashing the most terrible display of the power within the atom.

After the war, research resumed, and in 1947 a team at Bristol University were using a photographic emulsion technique developed by Cecil Powell, who had studied under Wilson, to record cosmic rays. These photographic plates were left up mountains or taken on aeroplane flights to capture more cosmic rays, then collected, developed and studied. They found another new particle, and this one matched exactly Yukawa's predicted pion. The pion was notable for a kinked track: sometimes it would spontaneously decay into a lighter particle (the muon in fact), changing direction at the same time. The result looked like two tracks joined together end to end.

Blackett was now in Manchester, where George Rochester and Clifford Butler took up cloud-chamber research and made another discovery: a new kind of V particle. Blackett had first found V tracks when a neutral particle (a photon) decayed into two charged particles (an electron and a positron). This new V particle must again be something with no electric charge, but this one was decaying to positive and negative pions. It was eventually named the kaon ('kay-on', given the symbol K), and it had to be heavy, as it decayed into pions. But unstable heavy particles should decay quickly, and strangely the kaon lived for a long time. The group at Caltech confirmed this discovery, and added another type of kaon, this one carrying electric charge. Things were starting to get confusing: what were all these new particles, and how many more were there going to be? It would require the invention of more new technology to answer those questions.

Accelerating science

The potential of atomic power for non-military use generated huge interest and funding for fundamental science after the Second World War. Large-scale research labs were set up in the 1950s, including the Joint Institute for Nuclear Research (JINR) in Russia, the Conseil Européen pour la Recherche Nucléaire (CERN) in Europe, and several new labs in the USA which to this day are run by the Department of Energy. Particle-physics research was to move away from small groups at universities studying cosmic rays to large groups based at these labs and using the latest pieces of equipment: particle accelerators.

Later in this book I will explain how particle accelerators work, but essentially they do just that: accelerate particles, then smash them together. This is a way to mimic what happens when a cosmic ray hits the upper atmosphere, but now a detector could be placed right next to that collision, capturing everything that was going on.

To go with these new particle accelerators producing the collisions, there was also a new type of particle detector to study the results. In 1952 the cloud chamber was superseded by the bubble chamber, invented by another of Carl Anderson's students, Donald Glaser. The idea was similar, but instead of ionisation causing condensation of vapour in air, it would cause the formation of bubbles in a fizzy liquid (beer was used in one of the prototypes). The advantage is as simple as liquid being denser than air: particles moving very quickly would fly though a cloud chamber without leaving a trace, but are much more likely to hit and ionise the densely packed atoms in the liquid of the bubble chamber. This new invention allowed the experiments to study very high-energy particles for the first time.

As particles could now be smashed together almost on demand, there was also an explosion of data. These new experiments were big – for example, the Gargamelle bubble chamber at CERN was 2 m across and 4.8 m long, and is today on public display there. When operational in the 1970s it produced over a million photographs of particle tracks, and sorting through them was intensive work. Batches would be sent out from the labs to collaborating universities, many of which hired teams of people to sort through them. At my own university, UCL, there was at one point a team of thirty photo scanners, mainly young students, sometimes working through the night to quickly process the latest results – students always get the most glamorous jobs. In modern experiments, the data is in the form of digital readouts from sensors rather than photographs, and much of the sifting and sorting is now handled by computers – because of both the complexity of the signals and the overwhelming number of collisions being produced. Over six thousand million million collisions were analysed to discover the Higgs boson; sorting these at the rate of 1 second each would take almost 200 million years.

The eightfold way

The new particle accelerators and more sensitive detectors quickly paid off as a whole host of new particles were discovered in the 1950s. At one point, the subatomic world had seemed under control, with just the electron, proton and neutron making up all matter. Then five, ten, twenty new particles turned up, and it was not just the muon, pion and kaon that needed explaining, but an entire zoo: the eta (η), rho (ρ), sigma (Σ), xi (Ξ), omega (Ω) . . . there was a danger of running out of Greek letters to label these things. Something had to be done.

By 1960 it was time for the theorists to try to make sense of everything. And two people independently had the idea organising the particle zoo by the properties of the particles: Yuval Ne'eman and Murray Gell-Mann. Because these 'particle periodic tables' contained eight particles, Gell-Mann named this the Eightfold Way, echoing the Buddhist Eightfold Path.

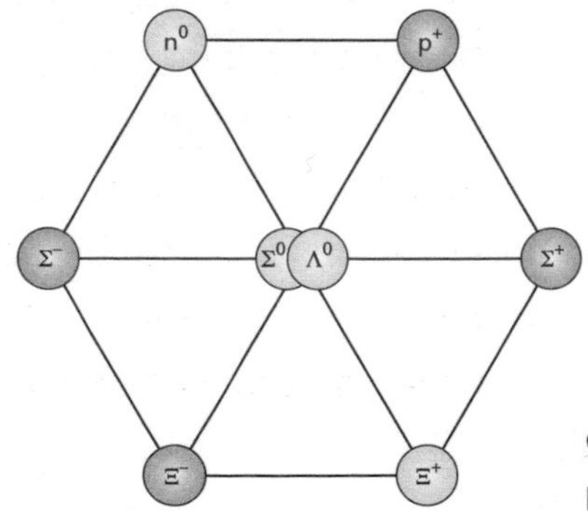

One of Gell-Mann's particle tables.

The patterns in these tables were a clue that there is some underlying cause, and by 1964 Gell-Mann realised what was going on. Everything could be explained using just three basic building blocks, which stuck together in various combinations to form the different particles that were being discovered in experiments. In another example of a simultaneous discovery, one of Richard Feynman's students, George Zweig, independently came up with the same idea. Zweig called these smaller particles 'aces', but it was Gell-Mann's name 'quarks' that stuck.

A word on pronunciation. Gell-Mann originally imagined something like the sound of a duck when coming up with 'quark'. Later, he found a line in *Finnegans Wake* by James Joyce: 'Three quarks for Muster Mark!', and now most people rhyme quark with Mark, not pork. Nowadays, for better or worse, things seem to be named with acronyms rather than literary references, which range from the mundane (LHC = Large Hadron Collider, which may one day be superseded by the VLHC, as in Very . . .) to the ridiculous (we have ATLAS = A Toroidal LHC ApparatuS; there is also WIMP = Weakly Interacting Massive Particle). Devising tenuous acronyms has become a favourite pastime amongst particle physicists.

Two of Gell-Mann's quarks are named 'up' and 'down', and make up the proton and neutron, as well as the pion and several other particles in the zoo. The proton contains three quarks: up,up,down. The neutron is down,down,up. To end up with the correct electric charge, the quarks must each carry a fractional amount: +⅔ for the up, -⅓ for the down, giving zero charge for the neutron, and +1 unit for the proton, which balances the -1 unit of the electron in the atom. Quarks also have antimatter partners, and like the antielectron, these antiquarks have the opposite electric charge to their matter partner. The pion contains one quark and one antiquark: up,antiup or down,antidown for the neutral pion; up,antidown or down,antiup for the charged pions. Particles containing three quarks are significantly heavier than those containing two, so were given the collective name 'baryons', from the Greek for heavy. Particles containing two quarks were given the collective name 'mesons', from the Greek for intermediate – they are lighter than baryons, but heavier than the electron. All particles made of quarks, both mesons and baryons, are named hadrons (from the Greek for 'thick', or 'massive') – as in the Large Hadron Collider.

To explain the kaon and other strangely long-lived particles, a third quark was needed: the 'strange'. This is heavier than the up or down, making the kaon (the positive kaon is made of up,antistrange) heavier than the pion (the positive pion is up,antidown). Some sense of order was returned: just the electron and three quarks were needed to explain the atom, and all of the particles in the zoo. Apart from the muon. Nobody knew what to do with that.

Building the Standard Model

A series of improved designs and new technologies made particle accelerators more and more powerful, producing increasingly high-energy collisions. The detectors studying these collisions also evolved, as bubble chambers were replaced by devices that could respond to collisions more quickly, and produce output that is easier to process than photographs: spark chambers, multi-wire proportional chambers, drift tubes, time-projection chambers, ring-imaging Cherenkov detectors, sampling calorimeters, silicon detectors and many more.

By the late 1960s, particle accelerators were powerful enough to smash apart protons, giving weight to the idea that they are made of smaller particles, the quarks. But there were still many doubts: nobody had ever measured a quark directly. Doubts increased when, as we shall see in chapter 5, the new theory known as the Standard Model suggested that quarks should always come in pairs. A new quark must exist – the lucky 'charm' to pair with the strange quark – just as the up quark paired with the down. But there was no sign of it.

Until 1974, when groups working at Brookhaven and Stanford Laboratories in the USA simultaneously discovered a new particle heavier than anything previously known, consistent with a meson made up of a charm and an anticharm quark. The Brookhaven group called this new particle the J, the Stanford group the ψ; today we use the awkward combination J/ψ. The Stanford group also found a heavier version of the same particle, so the J was dropped when naming that one the ψ, or 'psi prime'. The discovery of the charm quark triggered the 'November Revolution' in particle physics, sweeping aside all doubts about the existence of quarks and establishing the Standard Model as the correct theory.

By 1977, accelerators had advanced again, and the Fermi National Accelerator Laboratory (Fermilab) in the USA was leading the high-energy frontier. A new set of particles were discovered, hadrons that contained an even heavier quark, the 'bottom'. This must also come with a partner, the 'top' quark (the alternative names, 'beauty' and 'truth', fortunately never caught on). The top quark proved elusive until there came a machine powerful enough to discover it in 1995: the Tevatron, again at Fermilab. The Tevatron remained the highest-energy machine in the world until the LHC smashed that record in 2009. The LHC is almost 7 times more powerful than the Tevatron, and over 1,000 times more powerful than accelerators in the 1960s.

But what of the muon? Today we know this is basically a heavy electron – around 200 times heavier. It's one of the most common things you can see in the simple dry-ice demonstration cloud chamber, and is a well understood member of the Standard Model family. In 1975 an even heavier copy of the electron was discovered: the tau. Just as quarks come in pairs, the electron, muon and tau each also have a partner: a neutrino. Neutrinos are interesting enough to merit their own chapter later in this book, but for now these six particles complete the list of 'leptons', from the Greek for 'small', a name assigned before the heavy tau was discovered. And this completes the picture: 6 quarks and 6 leptons make up the 12 fermions, the matter particles of the Standard Model. A simple overview of this confusing multitude of particles is shown at the beginning of this book.

Big science

Particle physics started with a few individuals working at universities around the world: I picked out Cambridge and Caltech as a couple of hubs in the early twentieth century, but there are more. After the Second World War there was a shift towards larger, more powerful experiments based at national or international labs. There are certainly still individuals who make crucial contributions along the way, but particle physics at the frontiers has for some time been a team sport – and moves a lot faster as a result.

Today most researchers are, like me, based at universities around the world, while the experiments are based at the labs. The largest experiments on the LHC, ATLAS and CMS have over 3,000 authors listed on each scientific paper – meaning that over 3,000 people have made important contributions to the construction, operation, maintenance and data processing for those experiments. The team that actually performed the specific analysis in each paper is typically much smaller, anything from three to fifty or more – but no analysis would be possible without the contributions of all. Still, this is a lot of names, and one paper showing some combined ATLAS and CMS results has the dubious honour of holding the world record with over 5,000 authors.

The story I have told here has focused on the energy frontier, the most powerful particle accelerators which tend to make the big discoveries. But there is another side of the story, a wide range of smaller experiments that focus on measuring the properties of the particles once they have been discovered. These other experiments have told us just as much, if not even more, about the inner workings of the subatomic world, and I will mention some throughout the rest of this book.

The picture that emerged in the twentieth century, the Standard Model, is a huge success. It describes the universe at the highest energies and the smallest scales we have been able to measure. By studying particles in laboratories here on Earth, we learn how those same particles behave across the whole universe today, but also what happened less than a billionth of a second after the Big Bang. In fact, it is so successful that it is tempting to draw a parallel with classical physics – it works, but there are still a few details left over, and perhaps the next experimental discovery will be the one thread that pulls apart the whole picture once more.

Like most people starting out in physics research, I had a choice between working on experiments or theory. Theory has the big ideas, the beautiful maths and the powerful explanations, while experiments can be messy and frustrating. But I wanted to be part of that mess, part of the next big discovery – either a surprise like the muon, or a technological breakthrough like the cloud chamber that opens a new world for study. The LHC is looking for that loose thread at the energy frontier today, but it is not the only show in town. Later I'll talk about some of the many smaller experiments testing the Standard Model in different ways, all searching for the next breakthrough, the point at which the Standard Model breaks.

But first, there is more to learn about the Standard Model as it currently stands. It is not just a collection of fermions, it is really the fundamental forces that bring the world to life. So far I have only talked about the electromagnetic force, and the boson that transmits this force, the photon. We live in a world dominated by the electromagnetic force: the electricity that powers modern technology is the same as the force that holds electrons in orbit around atomic nuclei. Electrons in different atoms interact via the electromagnetic force, producing the solid world around us: the multitude of different materials, different colours, and the complex molecules involved in life itself. Sure, gravity holds us down on the Earth's surface, but the electromagnetic force makes it worthwhile to be here. Nonetheless the electromagnetic is just one of three forces in the Standard Model. The others behave in very different ways, but are also completely essential to the structure of the universe, as we shall see.

CHAPTER 4

THE UNDETECTABLE QUARK

For a short while in the 1920s, it seemed that the subatomic world might be quite simple: everything is made of atoms, and atoms are just electrons orbiting the nucleus, held in place by the electromagnetic force. This is enough to explain the periodic table of elements, and how atoms combine to form molecules and, in theory, very complex structures like DNA or the human brain. Of course, knowing the twelve notes in a scale does not mean you can play every piece of music in the world, just as knowing how atoms work does not mean we understand everything that can be built out of them, but certainly everything must obey the same basic rules, the rules of quantum mechanics.

But, as we have seen, the basic building blocks of the universe turned out not to be so simple. The discovery of the muon and the hadrons in the particle zoo meant that the rules needed to be revisited. Two other forces have to be added to the Standard Model: the strong and the weak. In some ways, they are similar to the electromagnetic force that we met in chapter 2: they are transmitted by bosons, which travel between particles that carry charge. But unlike the electromagnetic force, we don't experience either the strong or weak force directly. This chapter and the next one are about those two forces, why they are hidden from us, and how the world as we know it would certainly not exist without them.

Strong colours

The atomic nucleus was a mystery for a long time. First of all it's tiny – if the atom were the size of a sports stadium, the nucleus would be the size of a pea in the centre. It was only discovered in 1911 when Geiger and Marsden fired alpha radiation, high-energy particles, at a gold foil and saw some of them bouncing back. By the 1930s, the constituents of the nucleus, protons and neutrons, had been discovered, but what was holding it together? Protons all carry positive charge, and like all particles carrying the same charge they should be repelling, pushing each other apart and destroying the nucleus. Neutrons don't carry electric charge, so they can't cancel out that repulsion, but must be playing some role. To make a stable nucleus requires a delicate balance between the number of protons and neutrons – too many, or too few, and the atom became radioactive. The realisation that protons, neutrons and all the other hadrons in the particle zoo are made of quarks raised another question: what holds the quarks together to form hadrons like the proton and neutron in the first place?

It was clear there had to be some other force at play here, much more powerful than the electromagnetic. It was called the strong force, or strong nuclear force, and it is, appropriately, the strongest force that we know – but for a long time it wasn't clear how the strong force actually worked. Competing ideas were developed and discarded – Yukawa's theory predicting the existence of the pion in the 1930s was one of these. In the late 1960s, when particle accelerators were finally powerful enough to smash protons apart, it was possible to study this force in action, and a consensus started to appear. A theory called Quantum ChromoDynamics (QCD), based on the successful theory of the electromagnetic force, Quantum ElectroDynamics (QED), was able to describe all of the experimental data, and this is the theory of the strong force we use today.

There are many similarities between QED and QCD, but to understand why they behave so differently in practice requires a closer look at some of the details. In QED, the electromagnetic force is transmitted by a 'messenger particle' known as a boson: the photon. Any particle that carries electric charge can emit or absorb photons, and these interactions can be drawn using the simple Feynman Diagrams from chapter 2. By firing photons at each other, particles carrying the same electric charge repel each other, and particles carrying the opposite electric charge attract – these photons can act as 'balls' or 'boomerangs', to use the analogy from chapter 2. High-energy photons can also turn into a particle and an antiparticle, like an electron and positron, which carry opposite charge.

Now to extend this to QCD. It also has a force-carrying particle, a boson associated with it, and this is called the gluon (because it glues quarks together to form hadrons). Gluons can interact with any particle that carries the 'strong-force charge', which is named colour (hence the 'chromo' in QCD), though this is just a name, and has no more to do with real colours than some of the colourful language heard in heated discussions at the LHC. These interactions can also be represented by Feynman Diagrams: quarks are represented by straight lines like all fermions, and gluons are represented as curly lines. But this is where the similarities to QED end.

Photons are pretty outgoing. Particles can attract and repel by exchanging photons over huge distances, like a compass needle being moved by the magnetic core at the centre of the Earth. Other photons, such as light from distant galaxies, can cross billions of kilometres to reach us. But neither gluons nor quarks travel outside the comfort of a hadron – something like a millionth of a billionth of a metre. We have actually never seen a quark or a gluon directly, because they are always bound up like this, and we never see them in action until we smash hadrons apart. Something we generally don't do in everyday life, unless you happen to work at the LHC.

These differences between photons and gluons are due to

the nature of the 'colour charge'. Electric charge is a relatively simple quantity: it is 'one-dimensional', just like the number line. Electric charge can be positive or negative, and put two positively charged particles together and you get two units of charge; three positive particles give you three units of charge. Put a positive and negative particle together, and you get zero charge. And because opposite charges attract, positive and negative particles do tend to clump together to produce things like atoms which have zero total charge. Nature likes to sit at the symmetric point in the middle.

Colour charge is different. It is 'three-dimensional'. This means that as well as the charge having a positive or negative sign, it can also have a direction, and for convenience these different directions are labelled 'red', 'green' and 'blue' – 'up' and 'down' were already used to name the quarks themselves. So while a particle carrying one unit of electric charge is the same as any other, a particle carrying one unit of colour charge behaves like a weather vane, which can be pointing in one of three different directions: the red direction, the blue direction or the green direction. An up quark can be red, green or blue depending on how its colour charge is lined up – though this is just an analogy, quarks don't really carry a little pointer around with them. Antiquarks, the antimatter partner of quarks, carry opposite charges: for example an up quark carries +⅔ units of electric charge, and 1 unit of colour: red, blue or green; the antimatter partner to this quark carries -⅔ units of electric charge, and -1 unit of colour: 'antired', 'antiblue' or 'antigreen'.

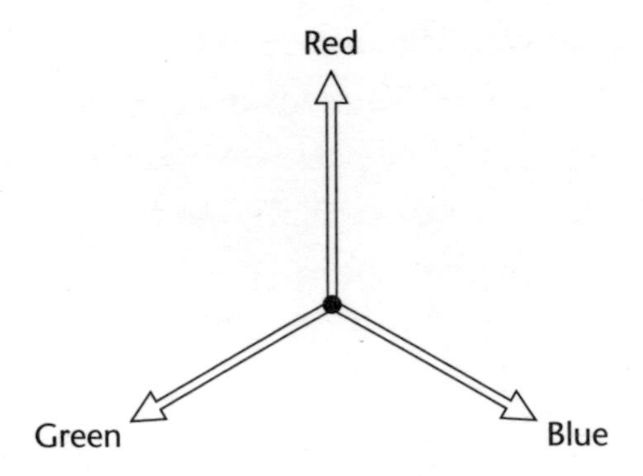

Representing the colour charge 'weather vane' on a quark.

Just as electrically charged particles tend to clump together in atoms with zero total charge, quarks always stick together in combinations that have zero total colour. A quark and antiquark pair can be colourless, for example red and antired makes zero total colour. These quark–antiquark states correspond to the mesons we met in the last chapter: pions, kaons, and so on. But it is also possible to stick together three quarks, one of each colour, and this is where the colour analogy works: red+green+blue = white, or colourless. These particles containing three quarks are the baryons, like protons and neutrons. The same holds for three antiquarks, which can combine to form things like the antiproton and antineutron.

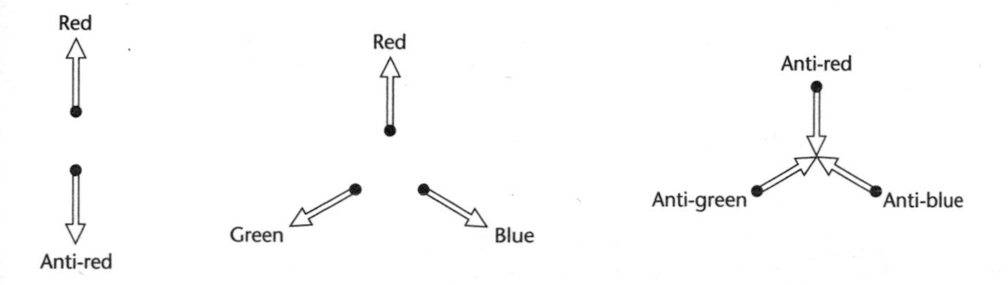

Colour charge alignment inside hadrons (a meson, a baryon and an antibaryon).

These are the simplest ways to stick quarks together to make colourless objects, but really there are limitless ways to do this, and more exotic hadrons have been discovered at particle accelerators. Tetraquarks contain two quarks and two antiquarks (e.g. red, blue, antired and antiblue), and have been seen in several experiments. Then in 2015 the LHCb experiment at CERN discovered pentaquarks, consisting of four quarks and an antiquark (e.g. red, green, blue, red and antired). The problem with with tetra- and pentaquarks (or indeed any more exotic combination) is that they can always be split up into two lighter combinations that are also colourless. For example, the tetraquark can split up into a red antired meson and a blue antiblue meson. The pentaquark can split into a red-green-blue baryon and a red antired meson. And this seems to be exactly what happens, and very quickly: the pentaquarks discovered at LHCb live for around 10^{-23} seconds, or a few millionths of a billionth of a billionth of a second. Any combination other than the basic quark–antiquark, three quarks or three antiquarks is hugely unstable.

Changing colours

This three-dimensional colour charge is central to the theory of QCD. Not only do quarks stick together to form colourless objects, but quarks themselves can change colour. And this is actually what the gluons do: if the quark colour charges are like little weather vanes pointing in one of three possible directions, the gluons are like the wind blowing them round: they don't change the total amount of colour charge, or the size of the vane, just the direction it is pointing.

This is only possible because of the crucial difference between photons and gluons: while photons do not carry electric charge, gluons do carry colour charge. Two types in fact: a colour and an anticolour. This is going to get confusing, but here is an example of how it works. A red quark can emit a red+antiblue gluon, and turn into a blue quark – the total amount of each colour always remains the same: red becomes red+blue+antiblue, which adds up to just red again, though now the colour is divided between the quark and gluon. This red+antiblue gluon may be absorbed by an antired antiquark, which will then turn into an antiblue quark: again, the total amount of each colour remains the same. So a meson made of a quark and an antiquark is constantly changing colour. It may start off red and antired, but after exchanging gluons these may switch to blue and antiblue, then green and antigreen, back to red and antired, and so on.

But you don't need to remember all that! The key thing is that no matter what colour the individual quarks are, things always remain in balance, and the hadron always ends up with zero overall colour. In fact, this constant switching of quark colours is completely invisible to us, as we cannot measure quarks directly. A hadron is like a swan serenely drifting along a smooth lake: we don't see the constant churning of colour within, the frantically paddling feet underneath the water. And this is constantly going on at the centre of every single atom inside every one of us. The world is a pretty strange place when you look closely.

Flying colours

With this picture of quarks as little weather vanes being blown in different directions by gluons, it's worth looking again at the comparison with QED and the photon. By digging into this, we see the reasons why the strong force behaves so differently from the electromagnetic force, along with the role the strong force has in shaping the world around us.

Photons interact with particles that carry electric charge; any electrically charged particle can gain energy by absorbing a photon, it can lose energy by emitting a photon, or can interact with another charged particle by exchanging a photon. The same applies to colour-charged particles and gluons, even though colour charge is apparently more complicated than electric charge. But now there is a big difference: photons themselves don't carry electric charge, so they don't interact with other photons. Gluons do carry colour, and this is what makes the strong force special: gluons can interact with other gluons. These gluon interactions require a new rule that can be used in building the theory, a new type of Feynman Diagram showing a vertex involving two gluons, and there is also one involving four gluons:

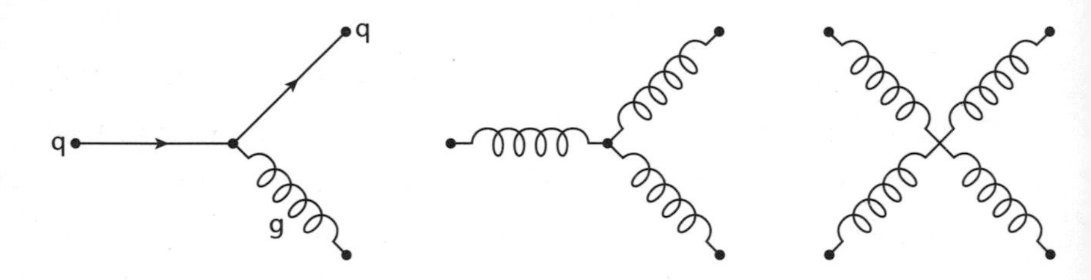

The components of QCD Feynman Diagrams.

As with the vertex in QED, these new gluon vertices can be rotated to any orientation. They tell us that one gluon can absorb another. Or that one gluon can split in two: gluons can multiply, and this is actually happening all the time inside hadrons.

It's something that we take for granted, that this doesn't happen to photons, that they simply pass right through each other. I have a 5 W LED light on my desk, and while this is not particularly bright, it still emits around a billion billion photons every second. Some of these photons travel straight to my eye. Some will light up the room, bouncing off the walls, off my desk, off my keyboard, and so on, scattering in all directions. These photons bouncing around are constantly crossing paths, but they pass straight through each other without noticing, and photons coming from any direction can reach my eye directly, despite having to fly through billions of other photons to get there.

If we could change QED by adding a rule in which photons carried electric charge and could interact with each other, the world would be completely different. We know how charged particles behave when they meet: they may repel each other, bouncing off each other and changing direction, or they may instead attract each other, sticking together. The same would be true of charged photons, and now one photon would even be able to absorb another photon, gaining energy as it did so. A photon leaving my lamp would then travel only a short distance before meeting another photon, and then it would either repel, changing direction, or absorb the other photon altogether. Then the same thing would happen with another photon, then another, and another, and so on and on. Imagine trying to run down a crowded commuter-train platform at rush hour, constantly bumping into other people and having to change direction, until someone completely blocks your path. The imaginary charged photons from my lamp would travel only a short distance before bouncing off other photons or being completely absorbed. Interactions between photons would mean that the electromagnetic force would be confined: I would no longer see my desk light, or indeed anything in my room, as no light would be reaching my eyes.

And this is exactly what happens to the strong force. We never experience it because the gluons carry colour charge, and can only travel a short distance before being confined. Everything is bound up tightly in hadrons, leaking out only a very short distance to stick hadrons together when they are right next to each other in the nucleus. Nothing carrying colour charge is allowed to escape, and only colourless hadrons are found in nature.

Because gluons carry colour, in theory there is actually another way to make a colourless particle: a glueball. As the name suggests, a glueball is a little ball of many gluons bound together, constantly dividing, merging and bouncing off each other, forming a particle with no quarks: all force and no matter. These strange things would be just as unstable as pentaquarks, and as yet there is not definitive proof that they exist, but they are another of the rich possibilities for exotic matter within the Standard Model.

Escape in a jet

At the Large Hadron Collider we smash protons together at high energies – more than high enough to break these protons apart. Understanding the strong force is the key to understanding almost everything that happens at the LHC, and in these collisions we get to see what happens over the very short distances that the gluons can act, and how the strong force behaves when provoked.

Before each collision, the quarks inside each proton are bound together by firing gluons at each other. When the collision actually happens, it is really a collision between two of those quarks, or even two of the gluons inside the protons. And the huge energy of a collision temporarily overwhelms the strong force: the quark or gluon that has been hit starts to fly out of the proton as if it is a free particle. But as it tries to escape, the strong force wakes up.

Let's say it is one of the quarks that is knocked out. The gluons that were flying between this quark and the others, binding them together, now have further to travel as one of them makes an escape. And the further they travel, the more likely they are to emit another gluon. Then another, and another, and so on. The strong force does not let go, but remains fairly constant as the quark moves away, as a stretching string of gluons reaches out to the escaping quark. Knocking a quark out of a proton is like kicking a ball up a hill – eventually it's going to run out of energy and roll back down as the gluons pull it back in.

This is different from what happens with the electromagnetic force. If I take my lamp and move it down the street, it appears dimmer – much too dim to illuminate my desk. Photons spread out in all directions, and the further they travel from a source like my lamp, the more spread-out they become. Because photons spread out in this way, the electromagnetic force also becomes weaker with distance (actually with distance squared). The strong force is completely different, and remains fairly constant with distance, because as the gluons travel further, they have more chance to multiply. Gluons don't spread out like photons, because they can always multiply to fill in any gaps. So as a quark tries to escape, the gluons continuously pull it back in. Eventually something has to give: either the quark will be pulled back into the proton, or if the collision was hard enough, this string of gluons will snap – a process known as 'hadronisation':

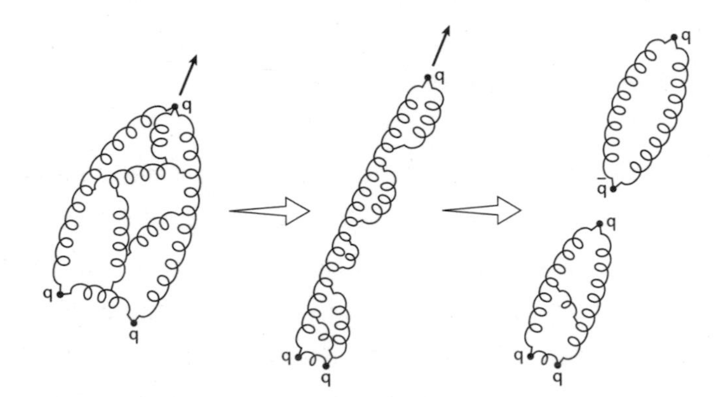

Gluon strings stretching as a proton breaks apart.

Once enough energy has built up in the string, the gluons can start making new particles – just as a photon with enough energy can spontaneously create a pair of charged particles in a matter + antimatter pair. One of the gluons in the string can spontaneously turn into a quark and antiquark, and this breaks the connection between the proton and the escaping quark. The new antiquark may now be bound up with this escaping quark to form a meson, while the new quark may be pulled back into the proton. Or, if there was enough energy in the initial collision, this process may repeat: quarks keep flying apart, more gluon strings snap, creating more and more hadrons. In collisions at the LHC, the protons are generally completely smashed apart, and we can see hundreds of new hadrons flying out. These new hadrons tend to form along the direction of an initial quark (or gluon) coming out of a collision, and they appear in collimated sprays known as *jets*. This is an important concept that will come up again in later chapters.

The $1,000,000 question

So far I've described how the strong force generally behaves, but what we really care about is using QCD to calculate things – like what can happen in a collision at the LHC. And what makes QCD really difficult to deal with in calculations is not just the extra rule that governs gluon interactions, it's that the strong force is so strong. In the language of Feynman Diagrams from chapter 2, this means that the 'coupling constant', the likelihood of extra gluons being produced, is large. The coupling constant for QED, α, is around 1/137, but for a collision at the LHC the strong coupling constant (α_s) is around ⅛. This means that the more complicated Feynman Diagrams with loops and extra lines are much more important in QCD than in QED – and tend to be fiendishly difficult to deal with.

One of the surprising things about QCD is that it actually becomes simpler at higher energies. The coupling constant isn't constant at all, but shrinks. Extra gluons become less likely. And at high enough energy, a quark could almost be knocked completely out of a hadron – a property known as 'asymptotic freedom', which was proved in the 1970s by David Gross, Frank Wilczek and David Politzer, and earned them the 2004 Nobel Prize in physics.

But there is a flip side to this: if the strong force is weaker at high energies, it must be stronger at lower energies. As quarks fly out of a collision, they start to emit gluons: α_s grows from ⅛ to ¼, ½ to approximately 1 – meaning there is almost a 100% probability for a gluon to be emitted, then another, then another, and so on. It becomes impossible to calculate things using Feynman Diagrams, and the models based on strings of gluons are used instead. Strings that stretch out between quarks snap if pulled too tightly, and eventually bind everything up in colourless hadrons. This is confinement, and the process of hadronisation.

We actually still have no rigorous mathematical proof that confinement follows from the rules of QCD, but it certainly seems to be what happens. The Clay Mathematics Institute based in the USA has chosen this as one of the seven 'Millennium Problems' and is offering $1,000,000 to anyone who can provide a proof of confinement, in what is known as the 'spectral gap problem'. It possibly became a little harder to claim this prize in late 2015 though, when the general spectral gap problem was shown to be 'undecidable' – a technical term meaning something like unprovable – and perhaps only in mathematics is it possible to prove that you cannot prove something! It's not yet clear if this means the specific case of confinement in QCD is also unprovable though, and right now the prize is still up for grabs.

This is a classic case in science: there is an 'easy' problem that we know how to solve (asymptotic freedom), and a 'hard' problem that we don't (confinement). There is an old joke about a dairy farmer who wants to increase milk production on her farm. She approaches some top academics from the local university for ideas, and after some time, the physics professor excitedly comes back. 'I've got the perfect solution! Though it only works for frictionless, spherical cows in a vacuum.' Not a classic, but the idea can ring true – it's tempting to turn any hard problem into an easy one by simplifying it as much as possible – but in the end, does the answer have any connection to the real world? This is where a lot of science happens: developing models, testing them, improving them, repeating.

QCD is a real cow. Calculations at the LHC involve stitching together two extremes: precise Feynman Diagram calculations for the initial part of a collision that we can solve, and gluon string models for the last part that we can't – along with some other models to splice the two extremes together. This has been a huge area of theoretical activity for many years, and one of the great successes in understanding the behaviour of the strong force. But because these calculations rely on models, they do need to be calibrated, to be 'tuned' to match reality. As experimental physicists, some of the first measurements we make are of basic things like the average number of hadrons being produced in collisions, and then we can tune the models to match the results. And after some tuning, the calculation can successfully describe the huge range of different things that can happen at the LHC. Essential really, because if we don't understand the basic physics going on in an average collision, we will have almost no hope of spotting the very rare collisions where an exotic new particle was produced somewhere in all the mess.

There has been progress using other techniques to model low-energy QCD, known as Lattice QCD. Instead of dividing a calculation into a series of more and more complicated Feynman Diagrams, Lattice QCD instead divides up space into a series of points, each interacting with its neighbour. These calculations are very intensive, requiring long processing times on large supercomputers, but produce results that are impossible to obtain any other way: predicting the masses and properties of various hadrons, as well as the existence of glue-balls and pentaquarks, for example. And as the calculations improve, it may be possible in the future to use them directly to produce even more realistic descriptions of the collisions at the LHC, with less need of the 'tuning' required today.

Life inside a proton

QCD also tells us a lot more about what goes on inside a hadron. Let's take the proton, the particle that we smash at the LHC. From a distance, this looks like a solid object. Look a bit closer, and we see the quarks: the simple model of a proton is three of them: up, up, down. But according to QCD, these quarks are constantly changing colour by exchanging gluons, like little levers constantly switching around in a well-balanced machine, and if we look closer we can pick out those gluons. And if we look closer still, we would see that some of these gluons are temporarily switching into short-lived quark–antiquark pairs, like the loops in Feynman Diagrams. The more we look, the more is happening, and collisions at the LHC are sensitive to this fine structure: they mostly involve gluons, but the collision may also knock out one of the main quarks, or even some of these short-lived quarks or antiquarks.

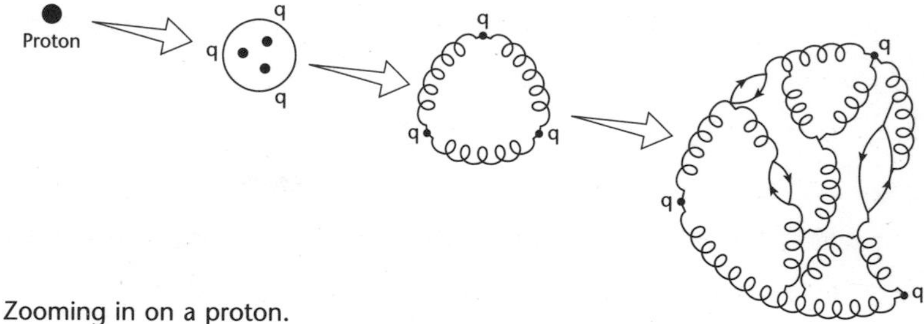

Zooming in on a proton.

There is a lot going on in a proton, and particles tend to behave in strange ways when squeezed into a small space like this, something first realised by Heisenberg in 1927 when he came up with the Uncertainty Principle – an idea that will come up again in more detail later in this book. Heisenberg wasn't worried about the strong force at the time, but he was trying to understand how electrons behave inside atoms. Classical physics said that they should just spiral straight into the nucleus, but this doesn't happen, and Heisenberg's answer to this problem was the Uncertainty Principle, which states that it is impossible to know both the position and momentum of a particle. As the electromagnetic force pulls an electron into a smaller orbit, it will pick up momentum and escape back out again. The electron orbits are this constant tug of war, like trying to squeeze a balloon only for it to bulge out somewhere else. If the electromagnetic force were a bit stronger, it could pull the electrons in closer – and as a result, atoms would be a lot smaller.

The same thing is happening to quarks inside the proton, but because the strong force is so strong, it can squeeze the quarks into a much smaller space: a proton is around 10,000 times smaller than the electron orbits in an atom. And because the quarks are confined to a much smaller space, the Uncertainty Principle says that they must pick up a lot of momentum – inside a proton the quarks are rushing around like caged animals, free to move until they push against the bars to try to escape, when the gluons pull them back in.

This constant high-speed motion inside all hadrons has another consequence, and a very important one for us. It follows from $E=mc^2$: Einstein's equation that tells us that energy and mass are equivalent, and indeed the energy of the motion of quarks and gluons inside a hadron appears as the mass of that hadron. More than 95% of the mass of protons and neutrons comes not from the mass of the quarks themselves, but from this 'binding energy', the zooming around of the quarks being held in a tiny space by the force of QCD. In other words, hadrons like the proton and neutron should really be light, but the energy required to contain the quarks inside such a tiny area makes them heavy.

Protons and neutrons are much heavier than electrons, and account for most of the mass in atoms. And the mass of atoms accounts for the mass of anything made of atoms: you, me, and everything else in the world. More than 95% of the mass of everything around us is due to QCD and this strange property of confinement. The other 5%? That's due to the Higgs boson, which we'll come to later in the book.

There are plenty of other strange things that emerge in QCD. While it usually collides protons, for a few weeks each year the LHC smashes together lead nuclei – still tiny on the grand scale of things, but these are really huge in terms of particle physics, containing 82 protons and over 120 neutrons. When smashed together, these collisions can temporarily create the highest temperatures ever produced in a laboratory – several trillion degrees Celsius – and an entirely different state of matter, in which the nuclei seem to melt into a soup known as a quark–gluon plasma, and the ALICE (A Large Ion Collider Experiment) experiment specialises in studying it at the LHC. This plasma would have filled the entire universe very soon after the Big Bang, when all matter was crushed together into a tiny area, and what comes out of the soup may tell us important things about how the universe evolved, and why it looks like it does today.

Hidden strength

This chapter has been a quick tour around the strong force, the force that holds the atomic nucleus together. This was a complete mystery for decades, but as experiments were able to start smashing protons apart, they showed that it is surprisingly dynamic and complicated. The theory of the strong force, QCD, started out as something fairly similar to QED: both describe a force transmitted by a boson, flying between particles carrying a charge. But adding one extra rule to QCD – the strong-force bosons also carry colour charge – gives the strong force a completely different character. And it is a character that changes: from asymptotic freedom at high energies to confinement at low energies.

The many different faces of QCD make it a fascinating subject, and I've spent a lot of my research career testing its predictions at the energy frontier. At the same time, we have the strange situation that all these predictions are based on things we can never see directly. Confinement means that quarks and gluons are always bound up inside hadrons, always wrapped up out of sight. QCD works like clockwork, but we can only see the face of the watch, never the wheels turning inside. So, if we cannot measure something, does it really exist?

This was one of the big questions in the early days of QCD, but now there is no real doubt. If those quarks and gluons didn't exist, collisions at the LHC would look very different, and we would need a completely different theory to describe them.

The sheer strength of the strong force makes calculations with this theory much more difficult though, so we rely instead upon a mix of precise Feynman Diagrams and models of gluon strings. But as with QED, the theory really works: QCD predicts the outcome of collisions at the LHC to high accuracy, along with many other aspects of the strong force. Confinement also tells us that most of the mass in the world around us is really due to the strong force squeezing quarks into the tiny proton and neutron, and that the insides of these hadrons are constantly changing. The strong force may seem inert most of the time, but take a peek inside the box and there is a whole world in there – and without this force, the universe would be a very different place.

The next chapter is about the other force in the Standard Model, one that is again similar yet different, and is equally important in forming the world around us.

CHAPTER 5

THE FORCE IS WEAK

With the strong force holding the nucleus together, and the electromagnetic force holding the electrons in orbit around the nucleus, we have the basic ingredients to explain pretty much all of the physical world around us. But there is one more force to add to the picture, and of all the things that happen in the world of subatomic particles, it is probably the strangest. This is the weak force, and although it appears to be very weak indeed, it turns out to be completely central to the Standard Model, and the construction of the universe. It allows the Sun to burn, and explains why everything is made just of protons, neutrons and electrons – without the weak force, we simply wouldn't be here.

Demystifying rays

The first clues to the existence of the weak force came with beta decay, one of the types of radiation classified by Rutherford in 1899, but as with the strong force, there were lots of different models developed along the way, and the weak force was not completely understood until the Standard Model was put together in 1968. And while this chapter is a summary of what we know now, rather than a history of how we got here, beta decay is one of the clearest examples of the weak force in action, so that is where I'll start.

There are four main kinds of radioactivity: alpha, beta, gamma and neutrons (Rutherford didn't know about neutrons at the time, so didn't call them 'delta'). These generally happen when an atomic nucleus is unbalanced, and it shifts into a more stable configuration; quite often a combination of these different kinds of radiation will happen one after another before a nucleus finally settles down.

First of all, the protons and neutrons may just reshuffle into a more tightly packed arrangement. This is a bit like packing a suitcase – the clothes fit a lot better if they are neatly folded instead of just heaped up. When a nucleus does some of this rearranging it releases energy in the form of a high-energy photon (a gamma ray). Sometimes reshuffling is not enough, and you can't close the suitcase at all without taking out that extra woolly jumper. In these cases, a chunk of the nucleus can break off, in the form of an alpha particle: two protons and two neutrons bound together. Or the nucleus may decide that two suitcases is the way to go, breaking up into two smaller nuclei, perhaps with some neutrons left over – the process of fission that is the engine of nuclear power stations.

But sometimes it's better to pack a completely different outfit, in which case you can also take out the tie that no longer matches. This is beta decay: the nucleus rearranges by swapping a neutron into a proton, and throwing out a beta particle: an electron. Beta decay can happen the other way round as well: a proton may change into a neutron, throwing out an antielectron – one of the processes that power the sun and all the stars in the universe. Either way, the combined number of protons and neutrons in the nucleus remains the same, but the balance between them changes, allowing the nucleus to become more stable.

The quark model tells us what must be going on in beta decay. The proton is made up of three quarks: up, up, down (for now we can ignore all the extra gluons, quarks and antiquarks that can pop up when protons are smashed apart at the LHC). The neutron is also made up of three quarks: up, down, down. So when a neutron turns into a proton, one of these down quarks must change into an up quark. Somewhere in there, the electron is also produced.

The problem is that there is no way to explain this using the theories we have met so far. The electromagnetic force allows electrically charged particles to interact, but they remain the same particle no matter how many photons fly around. The strong force can change the 'colour' of a quark, but the type of quark always remains the same. None of the forces or particle interactions we know can turn a down quark into an up quark, and there is no Feynman Diagram we can draw that can explain beta decay. To make this work, we need a new type of interaction, a new force. This is the weak force.

Like all forces, the weak is transmitted by a messenger particle, a boson. This is the W (yes, for Weak), and the W boson has some special properties: it can turn one particle into another. When a down quark emits a W boson, it turns into an up quark. Because the down quark carries -⅓ units of electric charge, and the up quark +⅔ units, the W must be carrying away -1 units of charge. The W must be electrically charged. It can then also decay into an electron, the beta particle that started all of this.

Precise measurements of beta decay showed that there was actually a problem with this picture: the electron does not seem to carry away all the energy released when the neutron turns into a proton. So in 1930 Wolfgang Pauli predicted that a new and unseen particle, the neutrino (actually an antineutrino here), was being produced along with the electron, and carrying away the rest of the energy. This gives us a much more balanced picture, which looks more like the Feynman Diagrams we have seen for QED and QCD: on one side, a particle emits a boson; that boson then turns into a particle and antiparticle pair (the neutrino is given the symbol ν).

We can build a full theory out of the weak-force Feynman Diagrams. These are similar to the Feynman Diagrams for QED, except that the particle type changes at the W vertex. There is one more vertex for the weak force: because the W carries electric charge, it can also interact with a photon. Because weak-force interactions are so rare that even the simplest possible Feynman Diagram gives a very accurate picture of what is going on – the chance of a more complicated weak interaction involving a couple of W bosons is so small it can be completely ignored. So this diagram tells us pretty much the full story of beta-decay:

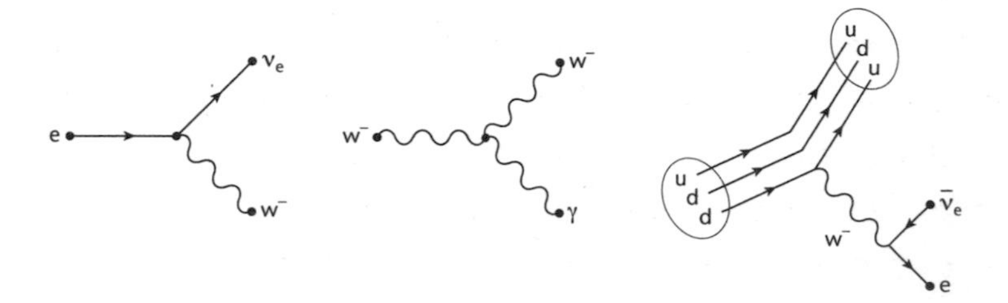

The components of weak force Feynman Diagrams, and on the right, the Feynman Diagram for beta decay.

We don't see the antineutrino in beta decay because it does not carry electric charge or colour charge, and only interacts through the weak force. This means it interacts so rarely that Pauli said: 'I have done a terrible thing, I have postulated a particle that cannot be detected.' Experimenters do enjoy a challenge though, and twenty-six years later Clyde Cowan and Frederick Reines achieved what Pauli thought impossible: measuring neutrinos. We shall return to neutrinos in chapter 8, as today they are one of the most active areas of particle physics research.

Why the weak force is weak

When I say the weak force is weak, I mean that interactions involving this force are rare. Beta decay can take seconds, hours, or even billions of years. However, one thing to know about the weak force is that it should not really be weak: all things being equal, it should be much more important than the electromagnetic interaction that is so crucial in our world: the coupling constant, α, of the weak force is about 5 times larger than the 1/137 of QED. But all things are not equal.

Beta decay is just one of the many kinds of particle decays that can happen, and there are two main factors that determine how quickly these decays take place. The first is simply how favourable it is. The neutron is just 0.1% heavier than a proton, so when a neutron turns into a proton during beta decay, some of that spare mass is turned into the lighter electron and antineutrino, and by $E=mc^2$, the rest is turned into energy: the lighter particles will fly away, like debris from an explosion. Many other particles can decay in similar ways, from muons to the heaviest of them all, top quarks. And the heavier the initial particle, and the lighter the final particles, the more favourable a reaction is. This can be thought about, believe it or not, as like a shop selling the latest top-of-the-range televisions. Normally, these very expensive models don't get bought very often. Offer a 20% discount, and sales probably pick up. Offer a 50% discount and they will go more quickly. At 90% off, they will be flying off the shelves. If a mid-range TV is also on sale, but at the same price as the top-of-the-range model, it won't sell as quickly. Decays work like this: the bigger the initial price (the mass of the initial particle) and the lower the sale price (the combined masses of the final particles), the faster a decay will happen. And beta decay is like the mid-range TV, with a smaller 'discount' and hence a rarer occurrence than some other types of particle decay.

While this discount is one factor, it is not the main reason the weak force appears weak. The big surprise about the W bosons is that, unlike the photon and gluon, the W has mass. It is actually very heavy: around 85 times heavier than a proton. To see how this affects a particle's decay, let's say the store is offering a 90% discount on the price of the television, but has to deliver it to your home. The price for this compulsory delivery is 85 times the original asking price of the television. Suddenly this isn't a bargain any more, and sales will stop pretty quickly. This is what happens in weak decays: the W boson is extremely heavy, and for a decay to happen, the energy to make this W boson has to be found somewhere. Looking again at the beta decay, we start with a neutron, then halfway through this process we have a proton, and an extremely heavy W boson. A lot of mass seems to have appeared out of nowhere. In a reaction like this, we might have enough money to pay for the television, the electron and neutrino at the end, but can't afford the delivery, the W boson. Beta decay should be impossible. The weak force shouldn't just be weak, it shouldn't happen at all!

Bending the laws of physics

There is one more piece of quantum mechanics that we need to include in our Feynman Diagram language, and we can see it in this case of beta decay. Looking again, there is a W boson in the middle of this diagram, but we can never actually measure it: it is emitted from a quark, then decays into an electron and an antineutrino. Particles like this, which are trapped in the middle of Feynman Diagrams, are called 'virtual particles'. We can never measure these virtual particles directly, and it turns out that they don't even have to obey the laws of physics.

There are a few really core rules in physics, rules that are respected by everything and tell us something deep about the universe itself. These rules generally say that certain things are conserved, that they don't change with time, and one of these is the conservation of energy. Basically you can't get energy for nothing. However much energy you start with, you have to end up with the same amount. Most of the things we do involve turning one form of energy into another: a kettle turns electrical power into heat; plants turn solar energy into sugars and other molecules that store energy; cars turn some of that chemical energy into kinetic energy (motion) by burning petrol. But in all of these examples, the total amount of energy remains the same. You can't boil a kettle without electricity, or run your car without fuel – energy must be put in to get energy out. This rule is respected by everything in the universe, from the behaviour of atoms to the motion of galaxies, and everything in between. By everything, that is, except virtual particles, the particles in Feynman Diagrams that we cannot measure.

This is another consequence of Heisenberg's Uncertainty Principle, which we met in the last chapter. It says that when particles are squeezed into a small space, they try to escape. But it also says that these virtual particles are allowed to bend the rules, to borrow a small amount of energy for a short period of time. And this is what allows beta decay to happen and the weak force to exist: the borrowed energy can be used to make the W, acting as a short-term loan covering the cost of delivery. But the more that is borrowed, the less likely it is to happen, and a huge amount of energy must be borrowed to make a W boson – this is what makes weak interactions rare. Borrowing such a large amount also means the loan gets called in more quickly, meaning that the W boson can only exist for a tiny fraction of a second before the energy has to be returned and everything balanced out. Because of this time limit, the weak force can only reach extremely short distances: around a billionth of a billionth of a metre. Smaller than an atom, than the nucleus, and even around a thousand times smaller than a proton.

These two factors (the discount on the price, and the loan needed for delivery) factor into the lifetimes of many unstable particles. The neutron lives on average for around 15 minutes – I say on average, because this decay is a quantum process, and there is an element of randomness in the time when any given neutron will decay. Inside a nucleus, the electromagnetic force affects the energy balance, and neutrons can become completely stable – a good thing for the atom! When a muon decays to an electron it releases much more energy, and so the average muon lives for only a couple of microseconds – still long enough to fly for many kilometres. Other heavy particles discovered in particle accelerators, like the tau lepton or hadrons containing charm and bottom quarks, release even more energy when they decay, and live for just nanoseconds, travelling only a few millimetres or less before decaying.

Particle pairs

The other thing we know about forces is that they are only felt by particles that carry the right charge. Quarks carry colour charge, so they feel the strong force. Almost all particles carry electric charge, so they feel the electromagnetic force. For the weak force there must be a corresponding 'weak charge', and actually all the fermions, all the quarks and leptons carry it. And just as the nature of the colour charge was so instrumental in determining the behaviour of the strong force, this weak charge also has a unique character.

In the previous chapter I pointed out that the electric charge is 'one-dimensional', while colour charge is 'three-dimensional'. A particle carrying one unit of electric charge is the same as any other, but a particle carrying one unit of colour charge can be pointing in one of three directions – 'red', 'green' or 'blue' – and gluons can change this colour, just like blowing a weather vane to point in different directions. The weak charge sits in between, being 'two-dimensional'. The weak charge has two possible states, like a coin. All fermions come in pairs, the heads and tails of the weak charge, and the W boson can turn one into the other, like flipping the coin.

From looking at beta decay, it is already possible to start grouping things into weak-charge pairs. The up and down quark must be one pair, as a W can switch between them during beta decay. From the other side of beta decay we know that the electron and neutrino must be another pair. All the particles we know sit in these pairs, so there must be three kinds of neutrino: the electron–neutrino (ν_e) partners with the electron, the muon–neutrino (ν_μ) with the muon, and the tau–neutrino (ν_τ) with the tau. The quarks also pair up: strange and charm, up and down, and so on. Heads or tails like a coin, with the W boson having the power to flip the coin over. These weak pairs give the last piece of structure to the Standard Model: so far the fermions have been grouped into quarks (which feel the strong force) and leptons (which don't), but they are also grouped into these weak pairs, two sides of the same coin.

This pairing was used to predict some particles before they were discovered. The strange quark was discovered in cosmic-ray experiments, and by 1970 when the weak-force model was developed it was realised it must have a partner, the charm quark, discovered four years later. The bottom quark was discovered in 1977, but it took eighteen more years before accelerators were powerful enough to make its partner, the top, the heaviest particle we know. Finally, the tau lepton was discovered in 1975, which also had to have a tau–neutrino partner. Because neutrinos are so hard to measure, this was not discovered until 2000. And that completes the picture. There are three generations of quark pairs and three of lepton pairs, giving the 12 particles in the Standard Model. And the W boson can turn any particle into its partner.

The weak force also explains why, given all the possible particles in the Standard Model, the world is made of just the lightest: the up and down quark, and the electron.

The simplest case in which to see this is in the decay of the muon. This can emit a (virtual) W and turn into a muon–neutrino. The W can then decay into an electron, and electron–antineutrino. After a few fractions of a second, there are no more muons. The same thing happens to taus, which emit a W, turning into a tau–neutrino, then the W will turn into either an electron and electron–antineutrino, or muon and muon–antineutrino. After a few nanoseconds, no more taus. Only the lightest leptons survive: the electron and lots of neutrinos – and only because there is nothing lighter they can decay into.

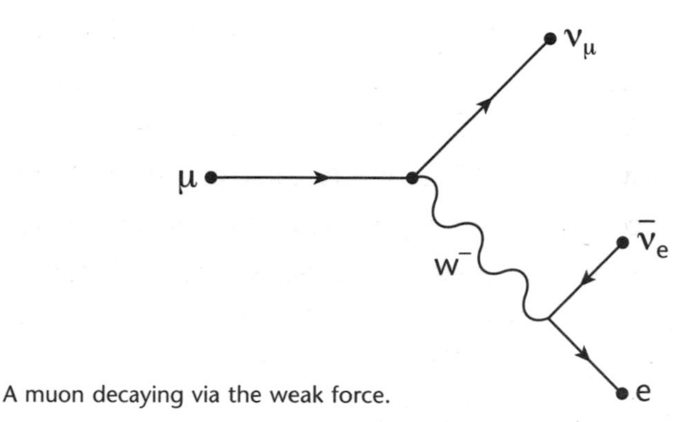

A muon decaying via the weak force.

Mixed-up quarks

We can try the same thing with the quarks. A heavy quark like the charm can emit a W and flip to its lighter partner, the strange. But the strange doesn't seem to have anywhere to go: the W can only flip it back into the heavier charm, which won't usually happen without some energy being added. In other words, the strange quark should be stable, because there is nothing lighter for it to decay into. The same is true for the bottom quark: this is the lighter partner of the top, and again it does not seem to have any way to decay. The world should be full of strange and bottom quarks. We should find them in atoms – though we don't, and the whole universe, would look very different if this was the case.

Clearly, there is something a bit more complicated going on for the quarks. While the leptons do fall exactly into these pairs, the quarks are a little bit mixed up. A W boson can turn a charm quark into a strange quark, but it can also turn it into a down quark or a bottom quark. In fact, the positive W can turn any positively charged quark (up, charm, top) into any negatively charged quark (down, strange, bottom); and the negative W can do the opposite. The amount of mixing – for example, how likely a charm quark is to flip into a strange quark compared with a down quark – can be measured by studying hadrons that decay by emitting a W boson. Experiments like LHCb at CERN (which also discovered the pentaquark mentioned in the last chapter) specialise in this kind of analysis, and the results are summarised by what is known as the CKM Matrix, after the three main developers of these ideas: Nicola Cabibbo, Makoto Kobayashi and Toshihide Maskawa.

It's not obvious why the quark interactions with the W get mixed up like this. The quarks do come in pairs, so the simplest thing would be for those pairs to line up perfectly as weak-charge pairs. At the moment, the best reason we can give for this is that it happens simply because it can. This mixing is technically possible, and so it happens in nature. It's not a huge effect, and the quark pairs do almost line up into W pairs: the top decays to the bottom quark 99.8% of the time, and a charm quark will decay to a strange 95% of the time. But the Standard Model does not provide any explanation for why these are not 100% – or even any other number. These are more 'free parameters', the numbers we can't yet explain, and the reasons they have the values they do remains one of the open questions in particle physics and I'll come back to them in chapter 9. However, if this mixing did not exist, the universe would look completely different – full of different kinds of matter that contain strange and bottom quarks.

Mixing means that the W boson can largely take its pick of which quarks it pairs up with in any interaction, and this means the heavier quarks can decay all the way down to the lightest: stepping down from top to bottom to charm to strange to up, for example. So mixing is a bit odd, but has played a very important role in the universe. Mixing does actually happen for leptons as well, but only appears in some of the strange behaviour of neutrinos that I'll talk about in chapter 8.

And so because of the weak force, all the heavy particles in the Standard Model are unstable. Very soon after the Big Bang, the universe was incredibly hot and dense, and must have been full of taus, top quarks, muons, and so on. But within a fraction of a second all of these would have decayed away – thanks to the weak force. Only the lightest particles survived: the up and the down quark, and the electron – and these, along with lots of neutrinos, make up the world around us. The quarks and electrons eventually combined to form atoms and, after a few billion years, us. We are made from the cold dust left after the Big Bang.

Anatomy of a collision

We now have all the tools in place to understand high-energy particle collisions at the LHC, which are similar to the collisions happening all the time when cosmic rays hit the Earth's atmosphere. One process that demonstrates almost everything is top-pair production, which we can first look at using the simplest Feynman Diagrams. There are a few ways this can happen, but this is the most common – and it should become clear reading the description of the muddled chain of events that follow in this collision that a Diagram can speak a thousand words!

Before the collision, two protons are racing towards each other at close to the speed of light. When the collision actually happens, one of the gluons that was holding each proton together may interact, fusing together to form a single, very energetic gluon. A lot of the kinetic energy of the collision goes into making this gluon, which then has enough energy to decay into a top quark and top antiquark – a combined mass around 185 times greater than the two protons at the start of the collision.

Top quarks and antiquarks are heavy. So heavy that they can spit out a W boson without having to use the uncertainty principle to borrow energy. Top quarks and antiquarks decay to bottom quarks and antiquarks in around 10^{-25} seconds: a millionth of a billionth of a billionth of a second – so fast that we can never see the top directly. So fast that the strong force does not have a chance to bind the top into a hadron. After these decays, coming out of the collision we now have a positively charged W, a bottom quark, a negatively charged W and a bottom antiquark.

The two W bosons also decay immediately. They can decay into any of the particle pairs: let's say one decays to a muon and a muon–antineutrino, the other takes its pick from the range of possible quark combinations and decays into a charm quark and a down antiquark. This is the core of the collision, the part that can be calculated using a simple Feynman Diagram like this:

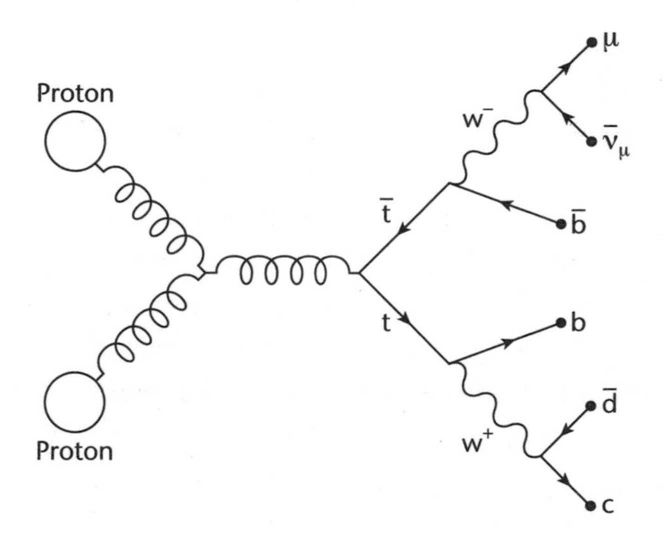

The basic Feynman Diagram for the first stage of top quark and antiquark production at the LHC. Before things get messy.

The muon and muon–antineutrino fly out of the collision, through our detector and out into the world. The muon will eventually decay into a muon–neutrino, electron and electron–antineutrino, but probably only after travelling a few kilometres.

For the quarks, the strong force takes over. They start emitting gluons, which stretch out into strings, some of which then snap, making more quark–antiquark pairs. Eventually all of these are bound up in hadrons in the process known as hadronisation. We had 4 quarks (bottom, antibottom, charm and antistrange), and so end up with 4 jets of hadrons.

The heavy quarks also decay. For example the bottom quark may be bound with an antiup quark in a hadron known as a B^-. The bottom quark decays to a charm quark by emitting a W boson after this hadron has travelled a few millimetres – far enough that we can measure it. This W can decay to more quarks, or to another muon and muon–antineutrino for example. More hadrons are produced.

The same thing happens to the hadrons containing the charm quark: the charm can emit a W and decay to a strange quark. Eventually, all that is left are the hadrons that are either stable or live long enough to travel a few metres into our detector. These are the pion, kaon, proton and neutron. There may also be some electrons and photons produced along the way; these too are stable and can be measured.

Finally, there are the remains of the two protons. This collision would completely smash them apart, sending quarks and gluons flying. The debris ends up producing many more hadrons, most of which carry on flying fairly close to the original direction of the beam. So the full collision looks a total mess:

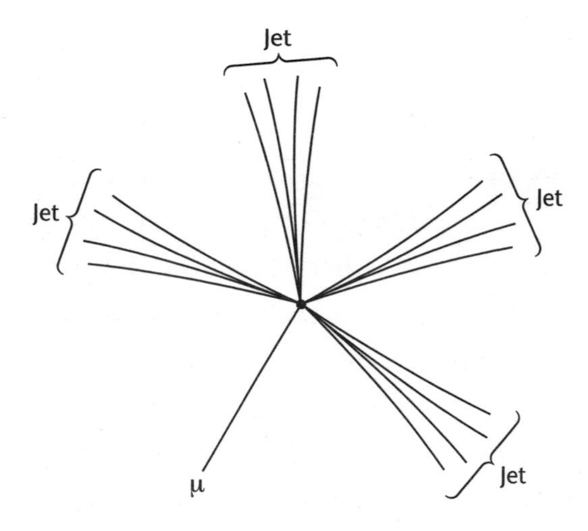

Some of the particles flying out of a collision following the decay of a top quark and an antiquark.

From all of this, we can pick out the muon and four jets, and reconstruct the top quarks that produced them. This is just one of the many many things that can happen in a collision, but because in the end there are only a few types of particle that actually live long enough for us to measure them, a lot of collisions can end up looking similar even if radically different things were going on unseen. Because we are colliding protons, jets of hadrons are the most common thing we see, and working out whether these came just from the two protons disintegrating or from something more interesting, like the decay of a top quark, is challenging. This kind of particle-detective work is one of the most enjoyable things about working on the LHC, or any particle-physics experiment, and I'll say more about it in chapter 7.

Hidden forces

The strong and the weak forces also provide the template for most ideas of what lies beyond the Standard Model. If there are more forces in the universe – and there almost certainly are – there must be a reason we have not already noticed them. Such a new force may be confined, like QCD: quarks, leptons or bosons may actually be made up of something even smaller, bound together by this conjectural new force. If we were able to collide particles hard enough we might overcome this force and start breaking them apart. Smashing open a W boson, for example. Or a new force might have a heavy boson, far heavier than the W or any other particle we know. Through the uncertainty principle we might be sensitive to this new boson in some very rare processes, just as beta decay was a clue to the existence of the W. Or if we put enough energy into a collision, we might be able to make this new boson directly. This happened for the W boson at CERN in 1986, when particle accelerators were finally powerful enough. Put enough energy into the collision and the uncertainty principle is no longer needed to make a W.

Weak interactions become more common, and collisions producing things like muons and muon–antineutrinos, electrons and electron–antineutrinos are detected in numbers.

But before going beyond the Standard Model, there are some loose ends to tie up. All the forces have something deeper in common, and that involves the final piece of the picture: the Higgs boson.

CHAPTER 6

THE HIGGS BOSON

The Standard Model works amazingly well. It describes the world around us in terms of tiny particles and their interactions. It also describes everything coming out of the high-energy collisions at the Large Hadron Collider, which means it describes almost everything that could possibly have happened in our universe since a tiny fraction of a second after the Big Bang. It does all this using a few different theories. QED describes the electromagnetic force and the photon. QCD describes the strong force and gluons. The weak force comes with the W boson and explains how particles decay. The Dirac Equation describes how the fermions, the matter particles, move.

But what is most exciting in all this is that all of these parts are connected. There is something much deeper going on in the universe, a guiding principle behind the Standard Model, and for what may lie beyond it: symmetry. Applying symmetries to our understanding of the subatomic world also showed that something was missing, and it led to the longest particle hunt in history: the search for the Higgs boson.

Symmetry

Symmetry is perhaps one of the single most important ideas in physics since the beginning of the twentieth century. It is the common link between Einstein's Relativity and the Standard Model of particle physics, the two great theories in physics today, and it may be the single most important thing shaping the whole universe.

When we think symmetry, we might think of something that looks identical when reflected in a mirror, and it is often said that people with more symmetric faces are more attractive. If we cut something down the middle equally, the two halves might be symmetric, or if we flip something around and it still looks the same, then we might say it is symmetric. As an example, a square has a lot of symmetry – there are several things that can be done to it, and after each one it will look identical: reflecting in a mirror; rotating by 90 degrees, 180 degrees or 270 degrees. A circle is even more symmetric: rotate it through any angle and it will look identical.

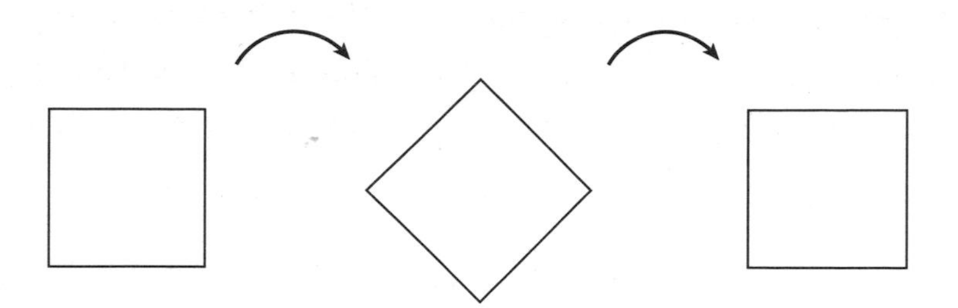

Symmetries of a square: rotate it by 90 degrees and it looks the same.

This can be turned into a more mathematical statement: a symmetry means that something remains unchanged after a transformation. Here, the 'something' can really be anything: someone's face, a square, or a subatomic particle. And a 'transformation' again can be any action: reflecting in a mirror, rotation by 90 degrees, changing the strong-force colour charge – really anything at all.

A look at the equations of physics that describe the world will show that there are symmetries there, and these symmetries tell us something deep about the universe. First, there are symmetries in space. 'Space' in physics really just means location: hitting a golf ball changes its position in space. And however you hit the golf ball, on any course anywhere in the universe, the laws of physics describing how the ball will move are the same. This is a symmetry: the laws of physics are independent of location, of transformations in space. Of course there are things that do change: each course has bunkers in different places, for example. But the underlying equations, the laws of physics, are the same, it is just the numbers that go into those equations – like how many bunkers there are – that vary from place to place. There is also a symmetry in time: the equations that describe how a ball will move are the same today as they were yesterday. Again, it might rain one day and not the next, but this just changes some numbers going into the equations, not the structure of the equations themselves.

These space and time symmetries can be collected together in what is known as the Poincaré Group, and they underpin Einstein's Special Theory of Relativity. Relativity tells us, essentially, that space looks the same everywhere, and though this may seem obvious, it is actually a result of these symmetries. The symmetries do lead to all kinds of counter-intuitive conclusions though, such as the equivalence of mass and energy: $E=mc^2$. Einstein extended these ideas to gravity and accelerated motion in the General Theory of Relativity, again using symmetry as a guide. And while Einstein may have been the first to fully recognise the role of symmetry in physics, it was also essential in understanding the world of fundamental particles, and that is what we are interested in here. The symmetries in particle physics are to do with strange quantum properties, so they are quite abstract, but they do reveal the true structure of the Standard Model. We are diving into the deep end here, so take a good breath!

Of particles and time zones

To tackle the first quantum symmetry, we have to go back to some of the ideas from chapter 2, specifically, how particles move: they split into multiple ghostly copies and explore all possible paths from A to B. In the example of my quantum journey to work, I get on the bus, which then splits into many different copies, spreading out and taking every possible route across town before coming back together when I arrive at work. Some of these routes can end up cancelling each other out, as when two buses arrive head-on in a narrow street and block each other. When everything happens all at once, not every path is possible.

There is a bit more to this, because if two buses are going to block each other, they have to arrive at the narrow street at the same time: arrive at different times, and they can both pass down the street with no problems. Particles must keep track of this timing somehow, and they do it by carrying something like an 'internal clock'. Mathematically, this is something called a *complex phase,* which works like a pedometer: a watch with just one hand that spins as a particle moves. The further a particle moves, the more it spins. At the start of the journey, before all the copies of the particle embark on their various paths, they all reset their pedometers so that the hands are pointing in the same direction. Then they all set off on different paths, some of which are longer than others, and so their pedometers will drift in and out of sync. When any two paths meet, the relative synchronisation of their pedometers tells us what will happen: if the hands point in the same direction, then they are 'in sync' and there is no problem. If the hands are out by 180 degrees, then this is equivalent to meeting head-on, and the two paths end up blocking each other. In between there is a partial blocking, a partial cancellation, and these paths can still contribute a little bit to how a particle moves.

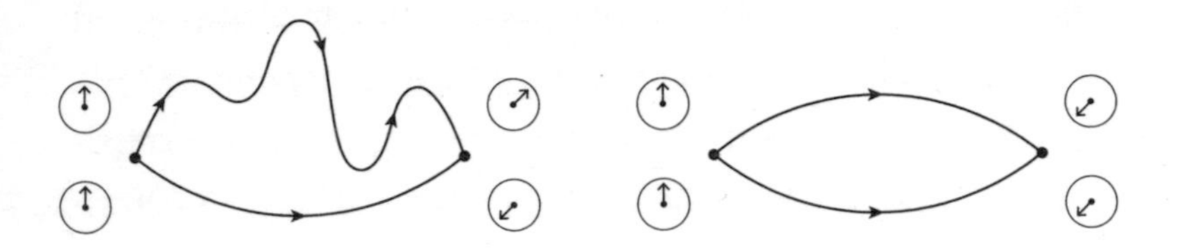

Some different particle paths that can end up being out of sync (on the left), or in sync (on the right).

Now, the symmetry: the behaviour of a particle is independent of the setting of its pedometer before it starts moving. The important thing is that all the strange quantum copies of the particle synchronise their pedometers to the same setting at the start. It doesn't actually matter if they line them up to point to 12 noon, or 3 p.m., or indeed any time at all – it only matters that they are all pointing in the same direction at the start, then drift in and out of sync as they move along their various paths. One of the consequences of this symmetry is actually that we can never measure the absolute setting of any particle's pedometer. This somehow seems quite efficient: because absolute setting is not important, we are not even allowed to know what it is! All that matters is the relative synchronisations along different paths, and we can work these out from how a particle moves, from which paths are getting blocked.

This is the case for one particle and its quantum copies. Now let's see what happens when there are two distinct particles that meet and interact, like two protons colliding in the LHC. They both carry internal clocks – and what matters is really the relative synchronisation of those clocks, just as when considering a single particle taking multiple paths. So what happens to our symmetry now? Are both particles still free to wind their pedometer to any setting at the start of the journey? On the one hand, the symmetry has to hold, because we cannot know how any particle is setting its pedometer. On the other, it can't hold, because if two particles set their pedometers differently it would change their relative synchronisation, which means they would behave differently: they might now not interact at all. Something is missing.

There is an everyday analogy that might help here, and it's one that still occasionally catches me out: time zones. I live in the UK, but many of my colleagues work at CERN, which is on Central European Time, one hour ahead of the UK. Some work in California, eight hours behind the UK. Meetings are usually scheduled on CERN time, and a lot of people like me connect via video conference from around the world. If a meeting is scheduled for 2 p.m. at CERN, I need to connect at 1 p.m. in London. Colleagues in California need to connect at 5 a.m. The point is that we all have different labels for the same time: 1 p.m. UK is the same time as 2 p.m. CERN, 5 a.m. California. Our individual clocks are all out of sync, just as different particles' pedometers can be, but we still manage to make it work and connect all at the right time.

There is actually no single definition of 2 p.m. It is just a label, and one that varies from country to country, and even through the year: some countries move their clocks by one hour over the winter for Daylight Saving Time – and the date this change happens also varies from place to place! The same goes for particles: they all have their own internal pedometers, but there is no single definition of how they should all be synchronised. My clock will read a different time from my colleague's clock in CERN or in California, and similarly there is no reason why a particle should have its internal pedometer synchronised with a particle on the other side of the universe, or the other side of the world, or even with another particle being accelerated in the opposite direction around the LHC.

The trick to make things work is communication. For example, I can send out an email saying 'Let's meet in an hour!' or 'Let's meet at 2 p.m. CERN time!', and we all know exactly what that means, because we all know how to translate that into our local time. So to get around the problem of different particles having different settings on their pedometers, there must be some way for them to communicate, to figure out their relative synchronisation.

And there is. It appears almost like magic. It is possible to take the Dirac Equation from chapter 2, which tells us how a fermion like the electron moves, then apply symmetry: the result cannot depend on how that fermion's internal pedometer is set. In order for this symmetry to be true, for the equation to return the same answers, an extra piece has to be added. This extra piece describes an interaction: a way for particles to communicate, to figure out their relative synchronisation. This interaction is actually the electromagnetic force, and particles communicate by exchanging photons, the boson of this force. The electromagnetic force is the direct result of a much deeper symmetry in the universe. This is how the Dirac Equation looks after applying symmetry:

$$i\gamma.\partial\psi = m\psi + A\psi$$

There is some confusing notation here: the γ in the Dirac Equation relates to how particles move. In Feynman Diagrams, the photon is given the symbol γ, but in the Dirac Equation it is A. So the new part of the equation contains both the particle, ψ, and the photon, A, and this term is the mathematical version of the vertex in the QED Feynman Diagrams.

There is one other consequence: the new term that appears in the Dirac Equation does not include mass for the photon. Adding it would unbalance things again and break the symmetry. So the symmetry not only tells us that the electromagnetic force must exist, but that the photon cannot have mass. All particles without mass travel at the natural speed limit of the universe, the speed limit that we know as the speed of light because that is how fast the massless photons travel!

So particles all over the universe have different local synchronisations of their pedometers, and these synchronisations can vary at different times and in different places. The electromagnetic force exists to smooth all this out, and particles communicate their local settings and stay in sync by exchanging photons, the boson of this force. Similarly, collaborators on the Large Hadron Collider are spread all over the world, with local clocks that vary depending on location and time of year. The World Wide Web was invented at CERN to allow everyone to communicate, and we stay in sync by exchanging emails, the 'boson' of the web, the force that makes an international collaboration work.

The unsung heroine of physics

Symmetries appear all over the place in physics. So far I have mentioned that the universe has symmetries in space and time, and Einstein's theory of Special Relativity follows from these. Then in the quantum world, particles behave the same regardless of how their internal pedometers are set, and as a result the electromagnetic force must exist, allowing particles to work out their relative synchronisation and preserve this symmetry.

These symmetries also have other implications, which follow from one of the most important mathematical results for modern physics: Nöther's Theorem. As a woman working in academia in the early twentieth century, Emmy Nöther faced institutional sexism. As a Jew working in Germany in the 1930s, she was driven to America by the Nazi purges just two years before her untimely death. In her obituary, Einstein named her a 'significant creative mathematical genius', primarily for her work in 1915, when she proved that wherever there is a symmetry, there is a conserved quantity: something must be remaining constant.

Nöther's Theorem can be applied to all the symmetries in physics. The fact that space looks the same everywhere actually leads to the conservation of momentum, and the fact that the universe looks the same in time leads to the conservation of energy. The fact that the laws of physics don't change means that a clock pendulum will keep swinging. Friction and wind resistance tend to break these symmetries and slow things down over time, but that's why it's easier to calculate things for frictionless cows in a vacuum, like the physics professor in chapter 4. But the two basic rules of physics – conservation of energy and conservation of momentum – follow directly from underlying symmetries in space and time.

But here we are interested in the quantum symmetries: the fact that the universe behaves the same way regardless of how particles' internal clocks are set. By Nöther's Theorem, this symmetry also leads to a conserved quantity, and that quantity is electric charge: the property that allows particles to interact with the photons that are preserving the symmetry in the first place. So not only does symmetry predict the electromagnetic interaction and massless photons, it also means that electric charge is conserved: the total amount of charge in the universe will not change. And we see this in the Feynman Diagrams where a chargeless photon can turn into a negative electron and positive antielectron: the total amount of charge is conserved – it remains the same (zero in this case) through this interaction.

Satnavs and coin tosses

Symmetries can be applied to all of the other forces in the Standard Model. In chapter 4, I described the strong force and the colour charge it is associated with. This charge is 'three-dimensional' – the electric charge is just a number, but the colour charge is a number and a direction. These directions are labelled 'red', 'green' and 'blue', and the colour charge is like a weather vane that can point in any of these directions.

But just as there is no universal synchronisation for particle pedometers, there is also no universal definition of which direction is red, blue or green. The directions are relative, and the definition of which direction is red can also be different at different times and places. For example, when giving someone driving directions, you might say they will pass a shop on the right – but it really depends on which way they are travelling. When driving back the opposite way along the road later on, the shop will be on the left. Directions are not universal, only relative to which way we are facing. And it's the same for these colour-charge labels: the direction that one quark calls red, another quark might call blue.

And just as for the particle pedometers, the colour charge of a quark is always hidden from us. There is no way to know if a quark is red or blue or green. And that's because the absolute definitions of these colour directions don't matter. There is a symmetry here: the way quarks behave cannot depend on which direction we label red, blue or green. All that matters for quarks is their relative colour. To form a hadron, a quark and an antiquark have to be opposite colours. We can call this red and antired, but that label is arbitrary; we might well call them blue and antiblue, but the hadron will behave in exactly the same way regardless. And because this label is arbitrary, and can change with time and place, quarks must have a means of communication to figure out their relative colours.

So again to make a symmetry work, an interaction has to appear, something like a satnav for colour directions, and that is produced by the strong force. And this comes with a boson that transmits the colour information between quarks: namely, the gluon. Just as for the photon, the symmetry also requires the gluon to be massless, and by Nöther's Theorem there is also a conserved quantity: the total colour charge must remain the same.

Finally the weak force: this is the same idea, only now

there are two 'dimensions' instead of three. In chapter 4 I described how particles come in pairs, which are like the heads and tails of a coin. Well again, the same logic applies: there is no universal absolute definition of what is 'heads' – all that really matters is the relative state that two particles are in. The universe should be unchanged regardless of how we define 'heads' and 'tails', and to make this symmetry work we need to introduce a new force, and a new boson: the W. There is a new conserved quantity to go along with this symmetry: the weak charge. Here the first sign of a problem appears: we cannot see a particle's internal pedometer, or the direction of a quark's colour weather vane. But the difference between the heads and tails of the weak force is very obvious: an electron is completely different from a neutrino. For some reason, this symmetry is not hidden from us.

Railway algebra

The symmetries behind the forces of the Standard Model are to do with labels that have no absolute meaning, definitions that can change in different times and places: the setting for a particle's internal pedometer, the direction of the colour-charge weather vane, or which way up we call heads or tails for the weak force. These types of symmetry are known as gauge symmetries, a historical term that derives from the gauge of a railway (the distance between the twin rails). The choice of gauge is fairly arbitrary, and a railway is still a railway regardless of what is used. But when buying trains it is best to communicate what gauge you are using, otherwise nothing will work! Following on from this, the bosons that communicate the information associated with these gauge-symmetry forces are known as 'gauge bosons'; they are the photon, the gluon, and the W.

There is an entire branch of mathematics devoted to the study of symmetries, and sometimes the Standard Model is written in terms of this mathematical notation as 'U(1) xSU(2)xSU(3)' – there is a short appendix on this if you would like to know a little more about what it means. But the most important thing here is that the maths revealed another problem. There should be three kinds of W boson: one carrying positive electric charge, one carrying negative electric charge, and one carrying no electric charge at all. But we only see the positive- and negative-charged ones in nature; the neutral one is missing.

So now we have three problems. First, the weak-charge state is not hidden from us, unlike the colour-charge state and the particle pedometers. Second, we seem to be missing a boson: there should be a third W with no electric charge. And finally, gauge bosons are required to be massless for the symmetry to work, which is true for the photon and the gluon, but the W bosons that we do see in nature are extremely heavy – more than 85 times heavier than a proton. There seems to be something different about the weak force, and at this point we either have to throw out the symmetry idea and start again, or find some tricks to make it work.

Adding weight to the problem

The first clue to solving the problems with the weak force came in 1964: a loophole that allows gauge bosons to have mass. This is one of those cases of simultaneous discovery: the mathematical tools were there, and the experimental results were pointing in the right direction, so several people independently made the same leap of imagination. Robert Brout and François Englert first set out the idea in August 1964, followed within a couple of months by Peter Higgs, then a few months later by Gerald Guralnik, Carl Hagen and Tom Kibble (all following earlier work by Philip Anderson, and independent later work by Gerardus t'Hooft, along with an even larger cast working with similar ideas). The loophole that they all found is sometimes known as the BEH (Brout, Englert & Higgs) Mechanism, the scientists the Nobel Prize panel selected for the 2013 award, though Robert Brout had unfortunately passed away by then. Peter Higgs himself has referred to it as the ABEGHHKt'H mechanism (Anderson, Brout, Englert, Guralnik, Hagen, Higgs, Kibble & t'Hooft)! Here I'll use what is mercifully the most common name: the Higgs mechanism.

This loophole to give gauge bosons mass requires a completely new force of nature, one unlike any we have met so far: while the other forces are the result of gauge symmetries, this force breaks those symmetries. The Higgs mechanism says that there is a force field extending throughout the entire universe, the Higgs field, and that it is full of energy.

To see why this is different, we can compare it with another force field, like gravity. Gravity only does something when there is mass around – we feel the pull of the Earth now, but going out into space, far away from any stars and planets, there would be no gravitational pull. In empty space, gravity is 'off'. But even in empty space, even when there is nothing else around, the Higgs field is always 'on'. There is always energy there, and particles can tap into the energy that it holds. It's not a perfect analogy, but this Higgs field is like molasses filling empty space. Try to stir a bowl of molasses and it will be very hard to move your spoon, just as if it were very heavy. And this interaction with the Higgs field, this sticking, makes particles appear to have mass. This analogy isn't perfect, because something moving through molasses will slow down and stop; a particle moving through the Higgs field will pick up mass but keep moving. The Higgs field is not like the old idea of the aether, because it does not have a preferred direction and is not really "at rest" anywhere, it is just a constant supply of energy.

If this Higgs field exists, then it is possible to make the weak force work as a gauge symmetry. The W boson is a real gauge boson, and is naturally massless. But at some point in the very early universe, the Higgs field 'turned on' and broke the weak-force gauge symmetry. Through interacting, or 'sticking' to the Higgs field, the W boson acquires some mass. This is the loophole: the mass of the W is not a property of the W boson itself – which is not allowed by the symmetry – but is a result of the interaction of the W boson with the Higgs field. Because the weak-force gauge symmetry is broken, the symmetry between the different states of the weak charge is broken also, and we can tell the difference between the electron and the neutrino, for example. It's like looking at a reflection in a cracked mirror – it's easy to spot the places where the symmetry breaks.

The final piece of the puzzle was the missing neutral W boson predicted by the gauge theory of the weak force. Working out where it went required another change to our idea of what electromagnetism really is, and the development of a new theory: the electroweak model.

Rewiring electromagnetism

Following the flurry of publications on the Higgs mechanism in 1964, this idea for giving mass to gauge bosons might have ended up as just a mathematical curiosity rather than something that actually exists in nature. But in 1968, Abdus Salam and Steven Weinberg realised that the Higgs mechanism could be combined with an earlier theory put together by Sheldon Glashow to give the structure we see in the Standard Model.

Earlier in this chapter I introduced the idea of gauge symmetries, which follow from basic properties of the universe: there is no common synchronisation for all particles, no single definition of colour-charge directions, and so on. The electromagnetic force does seem to obey a gauge symmetry, a symmetry related to winding a particle's internal pedometer forward and back. And while this symmetry does exist in the universe, it turns out that it's not exactly what we perceive as the electromagnetic force.

The actual winding symmetry produces a force that is similar to electromagnetism, and that force comes with a boson, the B. For Boson. The B behaves in similar ways to the photon, and is associated with a charge that is similar to the electric charge. But is not exactly the same thing. This is quite a subtle distinction: we may have thought that particles were using solar-powered pedometers, but they are actually battery-powered. The outcomes look alike, but there is a difference under the cover.

This does seem like an unnecessary complication. But now the Higgs field enters the game, giving mass to the W bosons. And another interesting thing happens: the missing neutral W boson (called the W^0) becomes tangled up with this B boson from the winding symmetry, giving two slightly different particles. Almost as if these particles were two lumps of dough: most of the B lump is mixed together with a little bit of the W^0 lump to make a new dough ball, the massless photon that we know. The remaining bits of the B and W^0 stick together to form another lump, a particle that becomes very heavy, like the charged W bosons. This is a particle we have not yet met: the Z boson – 'Z' because it has zero charge, but also because it is the final piece of the Standard Model story. The W^0 and the B never appear directly in nature, only in these two different mixes: the photon and the Z.

Because they both result from this mixing, the photon and the Z are closely related. They interact in almost the same ways, but with one difference: the Z can also interact with neutrinos. In the language of Feynman Diagrams, the Z is represented by a wavy line, just like the W and the photon.

But there is one important difference between the Z and the photon: the Z picks up mass from the Higgs field. It is even heavier than the W bosons we know, around 97 times heavier than a proton. And as for the photon, the particle that we are most familiar with, the particle of light that allows the human eye to see the world, this is actually a mix of two deeper things, the B and the W^0, tied together by the Higgs mechanism. The electromagnetic force is not after all a fundamental property of the universe, but emerges through this electroweak model.

You couldn't make this up.

Finishing the Standard Model

The electroweak theory written down in 1968 survives intact to this day, and explains two of the forces in the Standard Model. Adding in the strong force completes the picture, all based on the gauge symmetries that make it work. There are 12 matter particles, and 3 forces. The strong force is based on a symmetry associated with a charge that can have 3 settings: the 'red', 'green' and 'blue' colour states. The weak force is based on a symmetry associated with a charge that can have two settings:

the two particles in the weak pairs (electron and electron–neutrino, for example). Finally, there is a symmetry associated with a charge that can have just one setting, a simple number. The Higgs mechanism mixes up the B boson from this symmetry with the neutral W^0 boson from the weak-force symmetry, meaning that neither of these particles appears in the world, but we get two mixes of them instead: the photon and the Z. The charged W boson and the Z boson pick up masses from the Higgs field, while the photon remains massless.

The Higgs mechanism solves a particular problem with the weak-force bosons, and the electroweak model also says that all particles acquire mass in the same way: the electron, muon and tau, and the quarks. The Higgs mechanism is the key to them all.

But is any of this a description of real events? Following the proposal of the electroweak model, this stood as an open question for fifteen years. To make us take these symmetries of the universe seriously, some strange things had to happen: there must be this Higgs field filling the universe with energy. The photon actually is a strange mix of the B boson and the neutral W^0. There is also a 'heavy photon', the Z boson, that nobody had ever seen.

The problem with the Z is the same as with the W: the high mass makes interactions rare. There is a huge 'delivery cost' associated with the Z. This can be overcome using the uncertainty principle, but the size of the loan required means that the interaction only happens rarely. The same is true for the charged W bosons, but the W is unique: processes like beta decay can only happen if the W exists. But most interactions that are possible with the Z are also possible with a photon, but without the cost: spotting the tiny additional contribution of the Z is really like trying to find the needle in the haystack. There were some hints – so-called 'neutral current' interactions involving neutrinos that could not be explained using photons – but the definitive proof came in 1983. The Super Proton Synchrotron (SPS) accelerator at CERN reached high enough energy to actually produce the W and Z bosons in the laboratory. I'll mention this accelerator again in the next chapter, but by colliding protons and antiprotons at high enough energy, the delivery cost of the W and Z is covered, and suddenly interactions involving them become highly probable.

A good example is the discovery of the Z boson through the production of an electron and a positron, shown in the Feynman Diagram below. Here, the collision involves a quark from the proton and an antiquark from the antiproton. These annihilate to form a photon or a Z boson, denoted by the Z/γ line because we can't actually tell which one was produced. This Z or photon then decays into an electron and a positron:

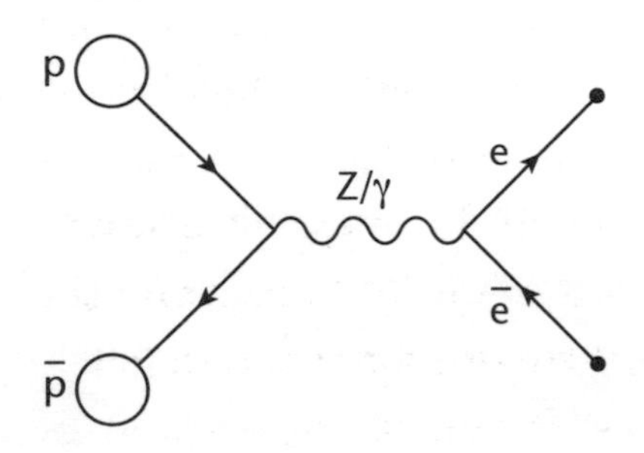

Producing electrons and positrons at the SPS.

By measuring the energy of the electron and positron that are produced, it is possible to work out how much energy was involved in the collision. And by $E=mc^2$, that energy corresponds to the mass of the virtual particle in the middle of the Feynman Diagram (individual Feynman diagrams are not the full story of what is physically happening, but this is still a reasonable approximation here). At low energies, the expensive Z is heavily suppressed, and the photon process dominates. The total rate of this type of interaction falls with energy, or with increasing mass of the virtual particle: we are less likely to find electrons and positrons at higher energy, as the virtual photon has to 'borrow' energy to make this happen.

But as the energy increases up to the mass of the Z boson, there is no loan needed. If the incoming quark and antiquark have enough energy, they don't have to borrow any more from the uncertainty principle to make the Z, and the rate of the process goes up. This is what was seen in 1983, proving one of the key predictions of the electroweak part of the Standard Model: the Z boson, this 'heavy photon', does exist.

At the LHC today, we can reach to much higher energies. If we see an electron and a positron with a combined mass far above that of the Z, it actually doesn't matter if the process involves a photon or a Z: they would both have to borrow some extra energy in order to reach this mass. As a result, they end up contributing more or less equally. At high enough energy, the symmetry between these two forces is restored.

And this is exactly what we see. We expect three regions: a

steady fall dominated by photons, a peak for the Z, and then a region where the photon and the Z contribute almost equally. And in all three, the theory matches the data perfectly:

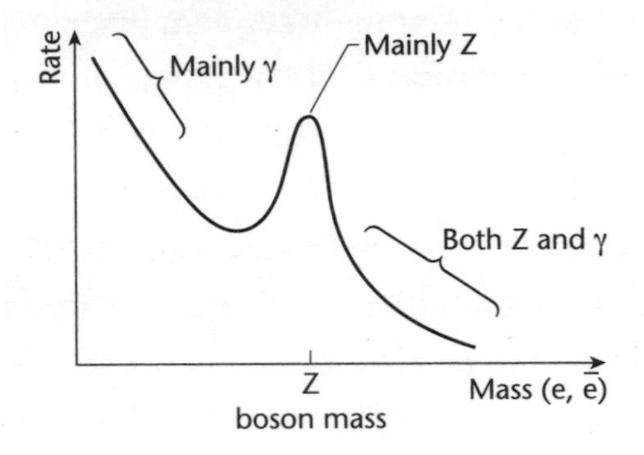

The rate of production at the LHC, at different combined electron + positron masses.

Higgs hunting

Even after the discovery of the Z boson, there was still the question of whether the Higgs field really existed, or whether something else was responsible for the 'electroweak symmetry breaking'. There were plenty of other ideas for how particle masses could be included in the Standard Model, and some direct evidence for the Higgs field itself was needed. Going back to 1964, Peter Higgs's paper was initially rejected by a journal that stated that it was 'of no obvious relevance to physics'. So he added one extra part and sent it off to a different journal – and this extra part was the first explicit mention of what the direct evidence could be, and this is probably why the whole idea is usually associated with his name. If the Higgs field does exist throughout the universe, it must come with an associated particle: the Higgs boson. This particle is equivalent to a ripple in the molasses, just as the photon is equivalent to a ripple in the electromagnetic field. Find the Higgs boson, and it would prove that the whole wild idea of gauge symmetries and the electroweak model was correct.

The Higgs boson would interact with any particle that has mass – these interactions are how particles pick up mass in the first place – so we can draw a set of Feynman Diagrams for the Higgs boson – and because the Higgs boson itself has mass, it must also have a 'self-interaction' vertex, where one Higgs boson can split into two:

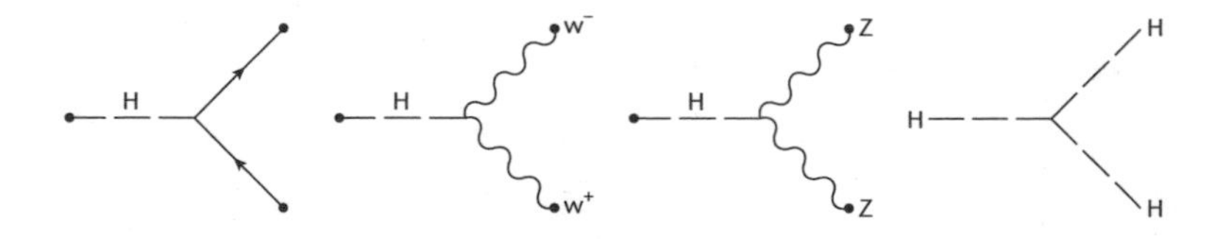

The components of Higgs boson Feynman Diagrams, from left to right: interacting with fermions, W bosons, Z bosons, and with other Higgs bosons.

These diagrams allow us to calculate how the Higgs boson would behave, and the kind of signal it might produce at the LHC. Unfortunately, while the theory told us that the Higgs boson should exist, and how it should behave, it gave very little idea of where to find it. It didn't tell us how heavy the Higgs boson is. Still with a powerful enough particle accelerator it should be possible to make it: essentially, to hit the Higgs field hard enough and make one of these ripples. But it was not clear how powerful that particle accelerator needed to be. Successive machines after 1983 went up and up in energy with no sign of it, and this became the last great mystery of the Standard Model: does the Higgs boson exist? If not, is the electroweak model incorrect? This prompted Nobel Prize winner Leon Lederman to name it the 'God Particle' in a 1993 popular science book, as if the Higgs boson were the key to everything. Or possibly this was the publisher's idea. The story goes that Lederman actually wanted to call it the 'God-damn particle', because it was proving so elusive; this story has a ring of truth to it, and no particle physicist today would call it the God particle! Yes, the Higgs boson is important, but there are bigger questions out there, as we will see in chapter 8.

The Large Hadron Collider was designed to run at such a high energy that it would finally settle this question. If the Higgs boson existed at all, it would show up there. If the Higgs didn't exist, the LHC would push the Standard Model to breaking point: the theory without the Higgs predicted that W bosons would start doing unphysical things, which really means something new had to turn up in the data – a new particle or a new interaction that would clean up the mess.

So when the LHC was turned on in 2010, after more than twenty years of planning and construction, there was a lot of excitement and much tension. And after just two years, the answer arrived: the Higgs boson was discovered by the ATLAS and CMS experiments, solving a 44-year-old puzzle, and finally completing the Standard Model – and I'll explain more about the discovery in the next chapter.

. . . *And relax*

There has been a lot of heavy lifting in this chapter. The Higgs boson is a special kind of creature; it has been the centre of a huge amount of fundamental research at CERN, and I'm always happily surprised by the number of people I meet who have heard of the LHC and the Higgs. But at the same time, understanding what it really does is not simple, even with a Ph.D. in particle physics. The fact is that the Standard Model is built out of some abstract gauge symmetries that the universe seems to obey, and all the forces are a consequence of those symmetries. That the universe is also filled with energy stored in this Higgs field. That some of the gauge bosons stick to this field, breaking a gauge symmetry and picking up mass. That all the fermions now have to do the same thing. And that the photon we know and love is actually a strange mix of the B and the W^0 bosons, two particles that we can never see individually.

On the other hand, there is an easy one-line summary of what the Higgs boson does: it gives mass to the other particles. And I have to admit that the catchy 'God particle' name definitely sticks in your mind. But understanding a little bit more about what the Higgs mechanism really is and how it fits into the picture opens a new window onto the strange, abstract and beautiful world of the subatomic particles.

And without the Higgs mechanism, the universe would be a completely different place. As chapter 4 explains, the proton would still keep most of its mass, as this comes through the effect of QCD squeezing quarks into a small space. But all the fundamental fermions would be massless – including the electron. Massless particles all constantly move at the speed of light, so electrons would be moving too quickly to be captured by atoms. All chemistry, and therefore all life as we know it, would not exist. And on top of that, the W boson would be massless, meaning there would be no suppression of the weak force: electrons would be constantly changing into neutrinos and back. It's hard to imagine the universe without the Higgs as anything but an empty place, just isolated particles zooming past each other in the cold dark night.

After three chapters of pretty heavy theoretical physics, it's time to get back to something more practical: how a machine like the LHC actually works, how the ATLAS and CMS experiments there actually found the Higgs boson, and are now looking for something completely new.

CHAPTER 7

THE BIGGEST EXPERIMENT

Geneva is tucked into the French-speaking corner of Switzerland, on the edge of Lac Leman (or Lake Geneva if your French is as good as mine), sitting between the Jura mountains to the north and the snowy peaks of the Alps to the south. Its austere cathedral harks back to the influence of John Calvin and the Protestant Reformation in the sixteenth century, while around 400 years later it became home to artists and thinkers fleeing the world wars. Now it hosts a mix of international companies and organisations, including the Red Cross and the UN. North-west of the city lies the suburb of Meyrin. Once a quaint village, Meyrin was bolstered by high-rise developments from that time in the middle of the twentieth century when concrete seemed to be the answer to every question. Walking on to the rhythm of clanking cow bells, next to fields of sunflowers in the summer, or under a permanent low grey cloud in the winter, we find CERN.

The 'Conseil Européen pour la Recherche Nucléaire', to give it its full title, was founded in 1954, partly in response to the Manhattan Project in the USA, and the realisation that answering the big questions in science now needed collaboration on an unprecedented scale. Twenty-one European countries are currently member states, with several others around the world involved to varying degrees. Thousands of people work at the main campus, and many more are based at universities and research labs around the world. Like Meyrin, CERN is something of a monument to brutalist architecture, proof that ideas can flourish in almost any environment. From the rather drab office buildings and warehouses, it is impossible to guess what is going on under your feet, or why there is such a buzz of intellectual energy. It's not until you visit one of the control rooms, packed wall to wall with screens displaying complex histograms and images, or one of the computing centres, huge rooms filled with racks of processors, that it becomes clear that unusual things are happening. And one of those things is the Large Hadron Collider, the world's most powerful particle accelerator, buried in a tunnel around 100 m underground and 27 km around. It extends out from the main campus, under the surrounding countryside, across the border to France and back. And having already discovered the Higgs boson, this might be the place that unlocks the next secret of the universe.

Smashing atoms

The LHC is the world's latest and greatest particle accelerator, the current pinnacle of technological advances starting from the early twentieth century. It accelerates protons to very close to the ultimate speed limit of the universe, the speed of light (99.999999% of the speed of light, in fact), then smashes them head-on into each other. The LHC is making these collisions at a higher energy than any previous experiment – almost 7 times higher than the previous record held by the Tevatron in the USA.

It is worth putting this record-breaking energy into context. When remembering how heavy a proton is in everyday terms (1.7×10^{-27} kg, or not very), then the huge energy going into collisions at the LHC becomes much less significant: it's roughly the energy of a mosquito in flight. The difference is that the LHC puts this energy into individual protons for every collision. Remembering how small a proton is (10^{-15} m, or extremely), this means the energy *density* is huge. For a tiny fraction of a second in each collision, the LHC packs a tiny amount of energy into an even tinier space, creating conditions that are unlike any other seen in a laboratory.

Why is this relevant? Looking at the universe today, it is expanding. Going back in the history of the cosmos, it must have been smaller. And around 13.8 billion years ago, it must have all been in one place, beginning with the Big Bang. We have no idea what happened at the Big Bang, but at that point the entire universe was contained in a vanishingly small space. It expanded rapidly, growing from nothing to the size of a grain of sand, the size of a football, the size of the Earth, the solar system, the Milky Way, all in the first few seconds. The universe continues to expand and cool, and the patch that we can see today is around 90 billion light years across. There is lots of space in between the planets, stars and galaxies, and the temperature of that space is now 2.7 K, or -270 °C.

Just after the Big Bang, the universe was a very different place. Taking all the galaxies and stars and everything else, and compressing it all into a space the size of a shoebox, it would have been extremely hot and dense. And this is the connection: the energy density created in the collisions at the LHC is equivalent to the energy density of the entire universe roughly a millionth of a millionth of a second after the Big Bang. The kind of particles we create and interactions we see at the LHC filled the entire universe at this time. Studying these collisions can tell us about the forces that shaped the universe into the form it has today, and about the origins of the particles that make up everything we now see, and others that we have yet to discover.

The LHC is probably most famous for discovering the Higgs boson. In the last chapter I went through the theory of the Higgs, and in this one I'll explain how the discovery was made: how a particle accelerator and detector works, and how the possibility of this new particle was turned into a reality.

So you want to recreate the early universe

To recreate the conditions of the very early universe, the LHC first needs protons. These are actually easy to come by: the most common chemical element in the universe is hydrogen, and hydrogen atoms are made up of one electron orbiting one proton; knock the electron off, and voilà. CERN uses just one small canister containing around 5 kg of hydrogen gas to supply the LHC. This might not seem like a lot, but then atoms are very small: 5 kg contains around $3x10^{27}$ hydrogen atoms. This is one of those ridiculously large numbers that it's hard to visualise, so to put it into context, if each of these atoms were the size of a football, they could be packed together forming a sphere roughly the size of the Earth. This canister contains enough hydrogen to keep the LHC supplied with protons for over a billion years.

To get the protons from this hydrogen, the electrons have to be removed, and this is done with an electric field, created by applying a voltage – like that from the mains power supply. To a particle that carries electric charge, sitting in an electric field is like sitting on top of a hill: positively charged particles will accelerate down one side of the hill (towards the negative voltage side), and negatively charged particles will accelerate down the other side (towards the positive voltage side) – this is just a simple way of describing the interactions of charged particles without having to calculate lots of Feynman Diagrams!

In a hydrogen atom, there is a negative electron orbiting a positive proton. If we put a hydrogen atom in a weak electric field, nothing much will happen: while the electron and proton would like to roll down opposite sides of the 'hill', they are also clinging to each other. Only when the 'hill' is steep enough (which means an electric field of a high enough voltage) will the proton and electron be pulled apart and go their separate ways – and a pretty steep hill is needed to split up a hydrogen atom into an electron and a proton. The first stage of the LHC is the spectacularly named 'duoplasmatron', which uses 90,000 volts to rip hydrogen apart (the process is a little more complex, as hydrogen atoms also stick together to form hydrogen molecules, H_2, but the basic idea is the same). This voltage is significantly more than the 110 or 220 volts in most domestic mains supplies, though it's probably a good thing that the voltages supplied to our houses can't rip atoms apart.

Once the protons have rolled down the 90,000-volt hill in the duoplasmatron, they will be travelling at speed, and this is a good opportunity to introduce the basic energy units we use in particle physics: eV, or electron volts. One eV is the amount of energy an electron would pick up if you accelerated it using 1 V (volt). Protons carry the same amount of charge (though opposite sign) as electrons, so pick up the same amount of energy in 1 V, though they are much heavier than electrons, so they end up moving a little slower. Either way, this is not a lot of energy – to put it into everyday terms, one calorie is around 4 hundred thousand billion billion eV, and there are 100 calories in a banana.

We also use 'natural units', redefining the speed of light, c=299,792,458 m/s, to be 1. This is equivalent to redefining the basic measure of length from the metre to the distance light travels in 1 second (roughly 3.3 nanometres), and has the advantage that $E=mc^2$ becomes simply E=m. Now mass (and even momentum) can also be expressed in terms of eV. For example, the mass of an electron is 511 keV, a proton is 938 MeV, and a W boson is 80 GeV (see appendix for an explanation of the prefixes).

The LHC is currently operating at energies of 13 TeV (13 trillion eV) by accelerating each proton to 6.5 TeV in opposite directions before colliding them head-on. Much less energy than a banana, but again it is the energy density that is important in particle collisions, squeezing this energy into a tiny space.

Some serious voltages will be needed to get the protons up to these energies.

Surfing through history

After the duoplasmatron, protons take a journey around CERN, a journey that is also a brief run through the history of particle accelerators. There are several stages, each of which was at one time the cutting-edge technology, but is now just adding a boost to the protons before passing them on. First is a linear accelerator, or linac. Linac 2, to be precise, which replaced, yes, Linac 1 in the late 1970s. Like all linacs, it accelerates particles in a straight line, and Linac 2 is just over 33 m long. A linac two miles long in Stanford, California claimed to be the 'straightest object in the world' and was the first particle accelerator powerful enough to smash protons apart into quarks back in 1968. Looking to the future, a new type of large linac is one of the possible ideas for what to build after the LHC.

Linacs, and their circular cousins synchrotrons, are the two basic designs of modern particle accelerators, and both use voltages to pull particles around: setting up a hill that the positively charged protons will roll down, speeding up as they go. Linac 2 accelerates protons to an energy of 50 MeV, 50 million electron volts. One way to achieve this is just to apply 50 million volts across the 33-metre length of the linac, turning it into one huge hill for the protons to roll down. But just as 90,000 V was enough to rip hydrogen apart, trying to put this kind of voltage across any piece of equipment would completely destroy it before the protons had a chance to do much.

The trick used to get around this is quite ingenious: instead of rolling down the hill of a fixed voltage, the protons ride along an oscillating voltage, just like surfers on an ocean wave. These oscillating electric fields are set up in RF cavities – the 'RF' means radio frequency, as the field inside oscillates at around the same frequency as radio waves, hundreds of millions of times a second. The cavities are essentially metal boxes, anything from 10 cm to over a metre in size. There are two holes: one for the protons to enter, and one on the other side for them to leave. Once protons enter the cavity, the wall behind them should be positively charged, and the wall in front negatively charged, so they receive a little pull of acceleration. Then as they leave the box, the charges will have flipped: the second wall now should be positively charged, so the protons are pushed away. The voltages flip back again when the next proton enters, and the process repeats. A whole series of RF cavities in a line will produce an accelerating voltage wave to accelerate the protons, and almost all particle accelerators today use them.

After surfing down Linac 2, the protons reach 50 MeV and are sent on to the next stages, and from here on they are all synchrotrons. These are circular in design, which has the advantage that protons will now pass through the same accelerating RF cavities over and over again as they travel around, receiving a little boost each time. In general, synchrotrons can push particles to higher energy in a more efficient way, but need to pull them around in a circle. This is done using magnetic fields.

While an electric field is like a hill that accelerates particles, magnetic fields are like banked corners, changing a particle's direction but not its speed. Back in chapter 3 I described how magnets were used in the early cloud-chamber experiments to bend positively and negatively charged particles in opposite directions, and the same thing is happening here. The electric fields in the RF cavities accelerate, and magnetic fields provided by large magnets steer the protons around in a circle. In reality, this is an incredibly delicate balancing act: the magnetic field must increase in strength as the particles increase in speed, as it requires a larger and larger force to keep them moving in the same circle. The electric fields in the RF cavities have to be synchronised within billionths of a second with the positions of the protons in order to deliver the acceleration at the right moment (hence the name 'synchrotron'). Computers are essential to making this work, but controlling the whole thing really is part of the art of running a particle accelerator.

Back to the journey of our protons, which after passing through the Proton Booster Synchrotron (PBS) enter the Proton Synchrotron (PS). This was built in 1959, and at that time was the most powerful accelerator in the world. It is 628 m in circumference, accelerates protons to 25 GeV, and was the accelerator used to provide the collisions studied by the Gargamelle bubble chamber I mentioned in chapter 3. From the PS we move to the SPS (Super Proton Synchrotron), another straight-to-the-point physics name. Built in 1976 and 7 km around, this can accelerate protons to 450 GeV. The SPS was never quite a record holder, as an accelerator at Fermilab in the USA was slightly more powerful at that time. But experiments on the SPS were the first to discover the W and Z bosons, a hugely important test of the electroweak theory built on the Higgs mechanism covered in the last chapter.

Finally, from the SPS, protons are moved to the LHC itself. This is a monster with a circumference of 27 km, and protons are packed into bunches and sent travelling in opposite directions around its ring. The LHC accelerates them all the way up to 6.5 TeV, and at this energy, protons make the round trip more than 10,000 times every second. Once loaded up, the bunches zoom around for some 12 hours. Colossally powerful magnets are used to keep the protons circling: 1,232 in total, each weighing around 35 tonnes, about 14.3 m long, and delivering a field of up to 8.3 Tesla. This is about 20,000 times stronger than a typical kitchen fridge magnet, and it would not really be practical to stick that much fridge poetry to the LHC. To deliver this kind of magnetic field in a controlled way, the magnets are superconducting, and have to be held at -271.25 °C, or just 1.9 degrees above absolute zero – colder than outer space.

To make the bunches collide, more magnets are used: the beams of protons are focused down to thinner than the width of a hair, then deflected slightly so that they cross paths at certain points around the ring: the precision needed to make the protons collide is something like firing two knitting needles from London and New York so precisely they collide halfway over the Atlantic. And at the points where these collisions happen (in the LHC, not over the Atlantic), the particle detectors ATLAS, CMS, ALICE and LHCb sit waiting to study the results.

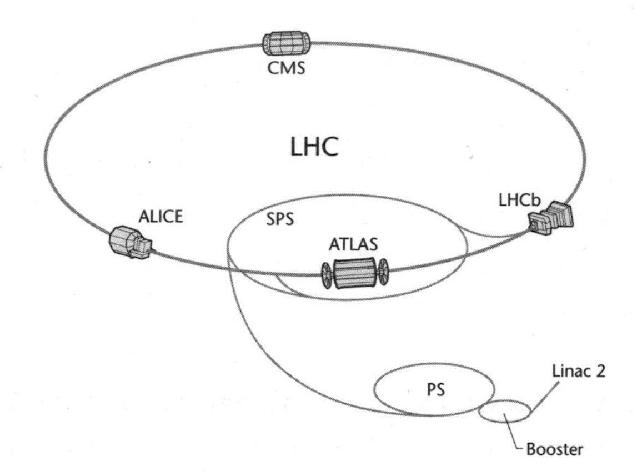

The CERN accelerator complex leading to the LHC.

That's what's going on inside the LHC, but what does it look like? The LHC sits about 100 m below ground, in a tunnel around 4 m high. It only takes up about a quarter of that tunnel, so there is plenty of room to stand up, and even ride a bike – the preferred method to get around the 27 km circumference when carrying out maintenance work. The protons travel round in a 'beam pipe', roughly 6 cm in diameter. There are two, one for the proton bunches travelling in each direction, and they have to be as empty as possible, otherwise protons will be lost in collisions with random air molecules as they travel around. So the pipes are held at a vacuum – emptier than outer space, in fact. The pipes are surrounded by the accelerating cavities and giant magnets, so if you were to go into the tunnel, these are what you'd see – interspersed with some of the cryogenics to cool the magnets, high-voltage equipment for the cavities and electronics to monitor everything. And all this requires a lot of power: when it is running, the LHC uses around 200 megawatts, much the same as a town of around 60,000 people.

Precision machines

At the start of this process we took a hydrogen atom and split it into protons and electrons. The LHC uses the protons, which are really a bag of quarks and gluons. It is impossible to know whether a quark or a gluon will get pulled out of the bag for each collision, or how much of the proton's total energy it will carry – it will certainly be less than the full 6.5 TeV. Electrons are fundamental particles; they carry all of their energy into every collision. So it's reasonable to wonder why the inefficient protons are being used instead of the electrons.

When particles are bent around corners, they lose some energy by emitting photons known as synchrotron radiation. Sticking with the analogy that our particles are surfing along in an electric field, synchrotron radiation is the water that is kicked up when the surfer makes a turn, and kicking up that water costs some energy. Lighter particles produce more synchrotron radiation than heavy particles. A lot more. The rate falls with the mass of the particle squared, squared. An electron is around 2,000 times lighter than a proton, and at some point it just becomes impossible to keep electrons going around in a circle at high speed due to the amount of energy they lose.

As an example, the LHC tunnel was previously used by the LEP accelerator, colliding electrons and their antimatter partner, positrons. LEP could reach an energy of just over 200 GeV, while the LHC, accelerating the much heavier protons around the same 27-km tunnel, reaches 70 times that energy. Some plans for a future electron–positron collider to reach higher energies therefore use a linac rather than a synchrotron: building a linac many kilometres long is still easier than dealing with synchrotron radiation. Even so, electron machines will never reach the energies of proton machines. Proton accelerators are generally 'discovery' machines: messy, but they push the energy frontier. Electron machines are 'precision' machines: they reach lower energy, but by knowing exactly what is going into each collision it is possible to study specific interactions in detail.

While synchrotron radiation is something of a problem in accelerators, it actually has a number of uses as well. Facilities such as the Diamond Light Source in the UK accelerate electrons, then deliberately send them around a series of tight corners to produce very pure, highly coherent beams of photons. These new light sources are being used to study the structure of materials to new levels of precision, just as Rosalind Franklin used X-rays to first reveal the double-helix structure of DNA. Synchrotron radiation will come up again in our discussion of particle detectors, where it is sometimes known as bremsstrahlung, a German construction translated as 'breaking radiation'. This is part of the international language of particle physics, a mix of words from English, German, French, Greek, Russian and others, along with some of our own inventions. One of the things I like about working at CERN, where there are so many nationalities, is overhearing snatches of conversation. A very excited and (to me) completely indecipherable discussion will suddenly make some sense when hearing the odd 'muon', 'NLO', 'calorimeter' or 'Higgs'.

Collision-scene investigation

Ultimately, a lot of collisions produced by the LHC look the same. One might produce a top quark and an antiquark, as

I described in chapter 5, or a Z/γ as described in chapter 6, or any of the thousands of other things that are possible – though because the LHC collides protons, most of the collisions simply produce a lot of quarks and gluons, bound up in new hadrons.

If any of the more exotic particles in the Standard Model are produced, they will decay long before they can be measured directly. After all these decays, the collision becomes just a spray of particles that do live for a reasonably long time, and the list of what these could be is a short one: electrons and muons; taus travel just a few millimetres before decaying, and can sometimes be identified from this flight distance. From all the many proveable hadrons, only the proton and neutron, pion, kaon, and a couple of others; some hadrons containing bottom or charm quarks also travel a few millimetres before decaying, so can be picked up like the tau. All of these hadrons are usually produced in the collimated jets I described in chapter 4. Finally, there is the photon. Neutrinos don't interact, so can't be measured directly, but their presence can be inferred – I'll come back to this in the next chapter. And that's it.

A single collision can result in hundreds of these detectable particles, and the aim is to piece these together, trying to reconstruct what could have happened. It's like finding a tiny fragment of pottery, then trying to work out if it came from a plate or a jug, or something much more interesting – for the archaeologist maybe a figurine, for the physicist a Higgs boson. It becomes easier the more fragments you find, so an experiment must cover the area around the collision to try to catch as many particles as possible. Then we need as much information about each fragment as possible: what kind of particle it was, how much energy it had, the direction it was moving, and so on. Any and all information helps in the reconstruction.

Most of the ATLAS detector, the experiment I work on, uses the same basic principle as Wilson's cloud chamber that I described in chapter 2, and that is ionisation: when an electrically charged particle flies near atoms, it is likely to knock some electrons loose. Using a voltage to sweep up those electrons (setting up another hill for them to roll down), they can be counted – by measuring the total charge collected, for example. It's also possible to count the photons emitted when electrons are recaptured by atoms. ATLAS does all this with around 100 million different sensors, making it something like a 100-megapixel digital camera – one that can take 40 million pictures every second. In 3D. Of subatomic particles. This is the kind of technology required to find the Higgs boson.

Not your typical digital camera

There are two main components in ATLAS: trackers, which measure particles as they pass, and calorimeters, which bring particles to a halt.

First come tracking detectors, which in modern experiments tend to be thin layers of silicon finely segmented into pixels, and arranged in concentric layers around the point where the collisions happen. As a charged particle flying out of a collision moves through these layers, it will produce a small signal in each pixel it hits. These signals in successive layers are like a trail of breadcrumbs, making it possible to reconstruct the track of a particle. To get the most information possible, the first layer sits very close to the collisions – roughly 3 cm away. The entire tracking detector sits inside a huge magnet shaped like a piece of rigatoni pasta 2.3 metres across and 5.3 metres long. Particles will curve in the magnetic field, just like the protons going around in the LHC, and measuring the bend along a track allows us to measure the speed (really, the momentum) of a charged particle in addition to its direction.

Outside the tracking detector magnet is the calorimeter, which measures the location and total energy of a particle by converting all of that energy into a pile of ionised electrons. Unfortunately, particles don't lose much energy ionising electrons, so they could fly for kilometres before coming to a stop, and we certainly don't have a detector big enough for that! To speed up the process, we need to make particles form a 'shower': turning one particle with a lot of energy into lots of particles with a little energy, little enough that they can be stopped in a reasonable distance. Calorimeters achieve this by alternating layers of detector with layers of a dense absorber, which can be anything from copper to iron, lead, or even depleted uranium. And it's this absorber that starts the shower.

The first particles that will be stopped by a calorimeter are electrons and photons. As these pass through the dense absorber, they interact with the electric fields of the atoms there: an electron may emit a high-energy photon, just like the synchrotron radiation in the LHC. A photon may convert into an electron and a positron. These may emit more photons, and so on. Looking like a firework, one initial particle quickly becomes an 'electromagnetic shower' of tens or hundreds of lower-energy photons, electrons and positrons that can be stopped and measured with just a few centimetres of detector.

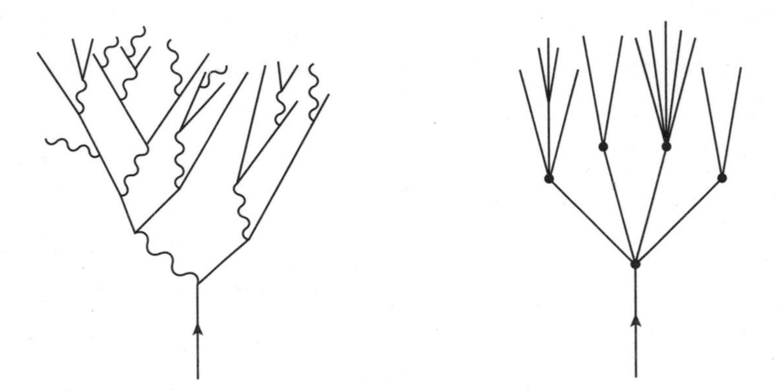

Electromagnetic (left) and hadronic (right) showers inside a calorimeter.

Any particle heavier than an electron is unlikely to emit photons – as with synchrotron radiation, the probability to radiate falls with a particle's mass squared, squared. All hadrons are too heavy to start electromagnetic showers, so must be stopped in a different way: hitting an atomic nucleus in the absorber to start a 'hadronic shower'. Such a collision causes the kind of strong-force interaction described in chapter 4: quarks and gluons are knocked loose, eventually being bound up in a handful of new hadrons. Because the nucleus is so small, these interactions are less common than the electromagnetic interactions of electrons and photons, and hadronic showers are harder to start. In order to try to catch them all, the ATLAS calorimeter is around 2.5 m thick.

This leaves the muons. Muons are around 200 times heavier than electrons, so they don't start an electromagnetic shower. They don't feel the strong force, so they don't start a hadronic shower. In other words, they are pretty unstoppable, so a second tracking detector is put outside the calorimeter to measure them as they fly by, zooming off into the world before eventually decaying far away. Muons produced in cosmic-ray collisions can even travel down through the atmosphere, down through the 100 m of rock, and be picked up by the LHC's detectors. The first thing I worked on at CERN in 2009 was actually to measure some of the properties of these cosmic-ray muons using the CMS detector before the LHC was even turned on – strange as it seems to use a detector buried underground in order to measure cosmic rays!

How different particles appear and are measured in a detector like ATLAS.

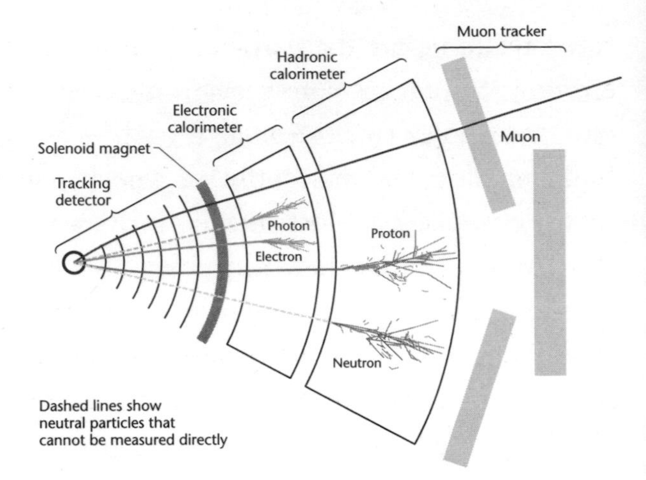

These are the components in any general-purpose particle detector: a tracking system surrounding the collision, with many layers of thin detectors to track particle paths without disturbing them. An electromagnetic calorimeter to start and capture the showers of electrons and photons, then a larger hadronic calorimeter to capture the showers of hadrons. Finally another tracker to record the muons as they leave. ATLAS combines all these elements in a roughly cylindrical detector standing over 25m tall and 45m long, whose different components tell us about the types of particles produced, as well as their energies and directions.

Collisions happen at the centre of ATLAS 40 million times every second; in other words, there is a 25 nanosecond gap between them. Particles flying out of a collision are almost always moving very close to the speed of light, but even at this speed will have travelled only around 7.5m in 25 nanoseconds: they won't even have made it out of the detector before the particles from the next collision are arriving. To prevent consecutive collisions just becoming a blur, the whole of ATLAS must be synchronised to incredibly high precision, and collect the information on each collision incredibly quickly. Most of the collisions are discarded at once using the triggers I mentioned in chapter 2: ultra-fast decisions based on signals seen in parts of the detector. Something like 1,000 of the most interesting-looking collisions each second are saved for later analysis.

Detectors with benefits

I've presented this as if it's obvious how to build a particle detector. It isn't, of course. Each successive design has been more precise, more sensitive and quicker than the last. They are both prototype and finished product in one. The technology is being invented as it is being used, from the magnets in the LHC to bend the proton beams, the fast electronics used to control the detector, through to the computing techniques used to process all the data.

These experiments have grown so complex that it takes literally thousands of people to make them work. But this is just a fraction of the true number of scientists, engineers, technicians, support staff and many others who have been involved since the planning began in 1992. This complexity, and the associated cost, means there is only one energy-frontier particle accelerator in the world at any one time. CERN estimate the total cost of the LHC to be around £4.2 billion, although this cost was spread between many countries over many years. For comparison, London paid about twice that to stage the Olympic Games in 2012. So yes, the LHC is expensive, but so are many large projects, and the LHC will be at the frontier of physics research for decades.

Still we are, quite rightly, often asked to justify the money spent to study things that are so unfathomably small, and seemingly so far removed from everyday life. I think that learning more about the universe, its origin and future and our place within it *is* the value that we get from such research. But the spin-off technologies that get invented along the way do also provide tangible benefits right now. Particle detectors have been common in medical imaging for some time, from X-rays to MRI and PET scans. Particle accelerators are now also being used to destroy cancer tumours at proton-therapy centres around the world. The specialised electronics, and of course new computing techniques – the World Wide Web and now grid computing – also find applications elsewhere. Working at the cutting edge of science necessarily involves developing both a deeper understanding of the universe, and the technology needed to test those ideas. It has been shown many times that investment in fundamental research like this returns a myriad of benefits; it is just not always possible to predict what will be discovered, or where those benefits will come.

Things to do with the world's most powerful particle accelerator

One of the first aims for the LHC was to discover (or completely rule out) the existence of the Higgs boson, thereby validating the electroweak theory at the heart of the Standard Model. The reason it took forty-four years from the development of the model to the discovery was the huge amount of technological progress needed to build an accelerator like the LHC, and a particle detector like ATLAS or CMS. But with these in place, the search was on.

From the last chapter, the Higgs boson sounds like a very strange thing: a ripple in a field that fills the entire universe with energy. But on the experimental side, we can take a much more pragmatic view of things: it's a particle, and we should be able to make it at the LHC. And if we can make it, it will decay into things we can measure. We just need to go and find the signature of these decays in all of the data.

The theory told us roughly how often the LHC would be able to produce a Higgs boson: about once in every 10 billion collisions. Finding that occasional collision really was like searching for a needle in a haystack – when you are not really sure if the needle even exists. To put this in proportion, the odds of winning the the main national lottery draw in the UK are about 1 in 45 million. If I bought one ticket for each of the two draws each week, I'd win on average once every 430,000 years. Producing a Higgs boson is 240 times less likely than winning the lottery, so colliding protons just twice a week is not really going to work – this is why the LHC does it 40 million times every second. And not just single protons, but bunches containing billions of them. This means that in each of the 40 million times these bunches cross each second, there can be anything from 1 to over 30 collisions. We call each of these bunch crossings, which may contain many collisions, an 'event'. Somewhere in all these events, all this mess, a Higgs boson will pop up every 4 minutes or so.

The theory also told us what the Higgs boson would decay into: any particle with mass. The actual discovery was made by looking for two different cases: the first where the Higgs decays into two Z bosons, which then decay into electrons and antielectrons or muons and antimuons; and the second where the Higgs decays into two photons. Here are the diagrams for these processes – and in case you were wondering how the massless photon can interact with the Higgs boson, it's via a loop:

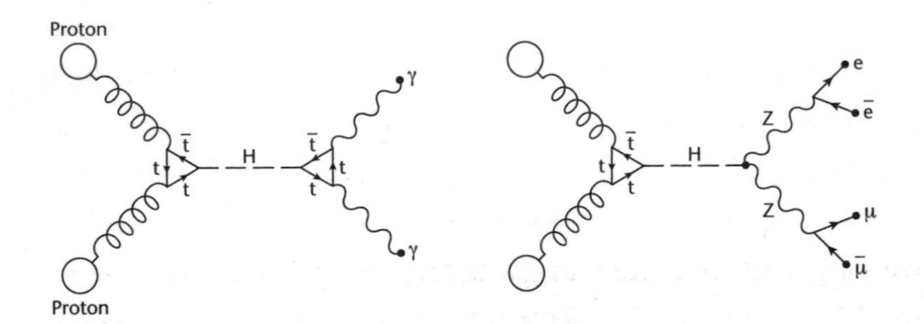

Feynman Diagrams for the Higgs signal at the LHC.

These two different decays produce two different signatures in the detector, and they were hunted down separately. I'll use the two-photon case as an example of how the search worked. First, the data was sifted, and all the events containing two photons picked out. And there were a lot, because there is more than one way to make two photons, for example when a quark and an antiquark emit one each:

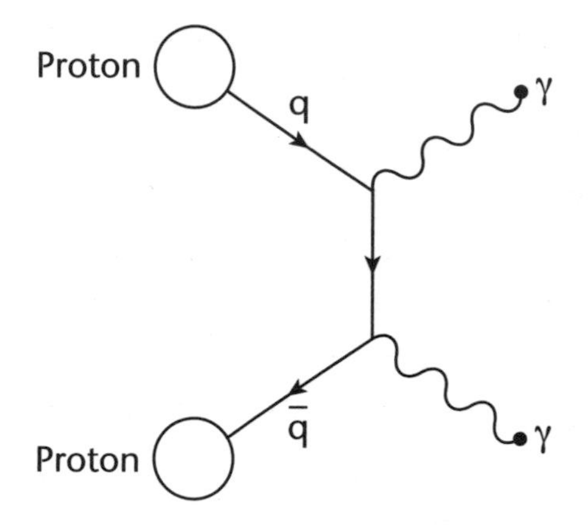

Feynman Diagram for one of the main backgrounds to the Higgs signal in the two-photon channel.

These things that mimic the Higgs boson signal are called backgrounds. Separating the signal from the background was the main task, and here there is one very useful fact: if we measure the photons precisely and combine their energies, we can work out the energy of the thing that made them. In the case of the background events, this will be a fairly random energy of two quarks pulled out of the proton.

Low-energy quarks are fairly common, high-energy quarks less so, but almost every collision will be different. In the case of the signal events, the two photons will always add up to the same thing: the mass of the Higgs boson.

The basic tool is then a histogram, which is just a series of pots that we call 'bins'. Say we have a histogram with five bins: 100–110 GeV, 110–120 GeV, 120–130 GeV, 130–140 GeV and 140–150 GeV. If we find a two-photon event, add the two photon energies and get 105 GeV, we add one counter to the first bin. Next two-photon event: 129 GeV, one counter in the third bin. 112 GeV: second bin, and so on. At the end, we just add up the counters in each bin. The background is more likely to give lower energy than higher, so the distribution of counters in each bin, starting with the 110–120 GeV one, might go something like: 101, 93, 78, 72, and finally 64 in the 140–150 GeV bin. Fewer counters in each bin as we go up in energy.

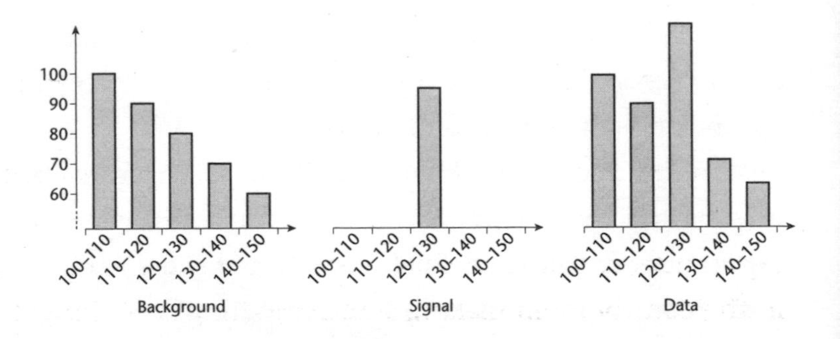

An example of how the background and signal may stack up in a histogram.

The Higgs boson signal – whatever it was – would all cluster in one bin. Say the Higgs boson mass was 125 GeV, every event containing a Higgs boson decaying to two photons will go in the 120–130 GeV bin. The number of signal events might look like: 0, 0, 42, 0, 0. In the data, the signal and background events come randomly through the year, so all we would see is the combined result: 101, 93, 120, 72, 64. There are more in the middle bin than we expect from backgound alone. A bump in the falling distribution, the clear sign of a discovery. Unfortunately, bumps are never this obvious in reality! There was also one thing the theory didn't tell us: the mass of the Higgs boson which gives the position where this bump would appear. So the net had to be cast as wide as possible.

Bump hunting

I quite often get asked what it is like to work on the LHC. Do we each get a time slot with the machine to run our own experiments? A slice of the data each? How can you be involved when you live in London?

When the LHC is delivering collisions, which is almost twenty-four hours a day for about six months a year, each experiment runs with a small crew (fewer than ten people) in the control room making sure everything is working: the data is being recorded, the detector has no problems, and so on. There are a number of experts on call in case anything goes wrong, but usually it's pretty smooth running. Usually a couple of times a day there is a break while the proton beams in the LHC are topped up. There are also longer breaks (a week or so) for technical work. During the winter, when electricity is too expensive to run the LHC, the detectors and tunnel are accessible for repair and development work – nobody is allowed down there while the LHC is running!

The data from the saved events is then processed. The latest calibrations and reconstruction techniques are applied, and rather than converting each event (which can contain many collisions happening at the same time) into a picture of what happened in the detector, it is converted into a list of objects and their properties: electron with energy of 45 GeV at an angle of 15 degrees, jet of 218 GeV at 75 degrees, muon of 15 GeV at 230 degrees, and so on. All of the processed events are then made available to everyone on the experiment to analyse, and we organise into groups interested in measuring specific things. Copies of the data are sent around the world to collaborating universities and institutes. And because the analysis is then all done on computers, it can be done anywhere: from CERN, from my office in London, or from the beach (with a good enough wifi connection). Being at CERN has definite advantages, and I visit regularly, because so many people are based there and it is easier to meet over coffee and get the latest information. But on each different analysis, there may be anything from a handful to over a hundred people involved, and they could be based almost anywhere in the world.

On a high-profile analysis like the two-photon Higgs search, a lot of people were involved (I was working on a different analysis at the time, but did play a role in delivering the processed data). They were not just sitting back and waiting for the data to arrive though. One of the main jobs of experimental physicists today is in designing new and better algorithms, tools and prescriptions for understanding and calibrating the detector, for sorting through mountains of data, and exploiting subtle differences between the expected signatures of the signal and background in order to make measurements and discoveries. And this is where the theoretical predictions – those precise calculations about what happens in an LHC collision that I mentioned back in chapter 4 – are essential. Without having a prediction for what to expect, it is very much harder to develop an algorithm to isolate any potentially new signal.

Day by day, we spend most of our time on this algorithm development and data analysis, sitting at the intersection of fundamental research and 'data science', and we have been dealing with 'big data' since long before that term became ubiquitous. We use a range of simple statistical techniques up to the latest machine-learning tools – using the theoretical predictions to train computers to classify events and separate signal from background.

In the Higgs search, the data was being processed quickly, and the search for the tell-tale bump continually updated. In order to avoid introducing any bias, all searches like this are 'blinded': the region where the potential signal may sit is removed from the data until the analysis techniques are finalised. Only then do we 'open the box' and see if a bump is there.

And after exhausting all the tricks of the analysis trade, it does come down to the numbers: collecting enough data that the bump would be obvious. Because collisions are quantum processes, we didn't know when exactly the Higgs bosons would be produced. Sometimes there is a clustering of background events that happens by chance, and produces something like a bump. This is a bit like tossing a coin and getting heads 5 times in a row: it can happen, and it doesn't mean the coin is special. Keep going, and the number of heads and tails will balance out. Collect more data, and these possible bumps tend to smooth out.

All except one. After collecting data in 2011 there were hints of a clustering of events at a mass of 125 GeV. This was well within the possible range for the Higgs boson, but not significant enough to claim anything. To fall back on the old science cliché: more data was needed. And after the winter break, this came and was analysed quickly. By 4 July 2012, both ATLAS and CMS were ready to reveal what they found. A seminar was planned for 9 a.m. at CERN (8 a.m. UK, midnight California), to be streamed around the world. People camped outside the auditorium overnight in order to get in. Most people, like me, watched it via video link in one of the overflow rooms.

Statistics, damn statistics, and the Higgs

Sorting the real bumps from the fakes ones requires the power of statistics. First we need to know how many events should fall in a given place. Here the theoretical predictions for what to expect from the backgrounds are essential – it is much harder to spot something unusual if we don't know what counts as normal. I find myself often thinking this when politicians or journalists quote statistics without giving the context: 'New study claims thirty people will have a heart attack this month!' Is that a lot? How many people normally have heart attacks? We need to know what to expect in order to make sense of this number.

Then we have to take into account the randomness of quantum mechanics. Say we take one part of the data, and make the histogram I described earlier. The number of events in each bin might come out like this: 101, 93, 86, 72, 64. Take another slice of data, and we might get: 103, 89, 85, 73, 62. We don't always get the same number in each bin, and we don't always get exactly the number we expect. Taking many slices of data will probably produce a slightly different result each time – there is nothing unusual happening, it's just that each collision is random. Toss a coin 10 times and you might get 6 heads; do it again and you might get 3 – it's the same effect. So we assign a 'statistical uncertainty' to each bin, which reflects the fact that we could have easily measured something a little bit higher or lower, purely down to chance. This uncertainty is chosen to cover 68% of the range of values we might expect due to randomness, and is given the symbol σ (sigma).

Then there is a systematic uncertainty which adds to this σ, increasing the possible range we could reasonably expect the data to fall in. Evaluating this is usually what takes 90% of the time in any analysis, and it can cover a whole lot of possible things. The detector can never be calibrated completely perfectly, so there is always a small uncertainty in the measured photon energies. If we use a different theoretical model, we might get a slightly different prediction for how many events our algorithms should select. And so on. The webcast of the seminars from 4 July 2012 is still available online, and you can see a lot of time is spent discussing the details of these systematic uncertainties, and how they affect the bump hunt.

Finally, we have an expected number of background events, and we have the actual number of events seen in the data. If the actual number in one or more bins lies 5σ above the expectation, 5 times the combined statistical and systematic uncertainties, then something interesting has happened. There is only a 1 in 3.5 million chance that the background could randomly produce such a huge shift – equivalent to a coin landing on heads 22 times in a row – so this is the threshold we set in order to claim that the bump is real.

And this is what the ATLAS and CMS experiments announced on 4 July 2012. Each experiment had been looking for the signature of the Higgs boson in events containing two photons, and also in events containing two Z bosons. And a bump showed up in both of these types of events – and at the same mass, around 125 GeV. Combining the two bumps crossed the magic 5σ threshold. The two experiments had worked independently up to this time, so that they could independently verify any result – and they both had a 5σ bump at the same mass: 125 GeV. The Higgs boson had been found at last, and the auditorium burst into applause.

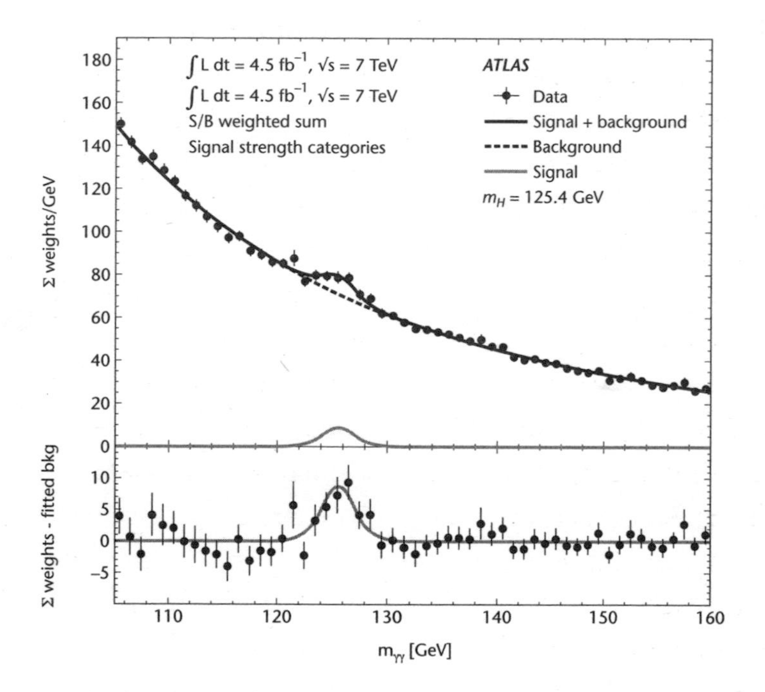

The Higgs bump in two-photon events from the collisions recorded by ATLAS.

The discovery completed the Standard Model. In 2013, the LHC was turned off for two years of upgrade work. Up to that point, it had been running at just over half power (7 TeV in 2010–11, 8 TeV in 2012) as a precaution: in 2008, when the LHC was turned on for the first time, one of the superconducting magnets overheated. The liquid helium used to keep the magnet cool boiled off, expanding rapidly and causing severe damage that took a year to repair.

In 2015, the upgrades were completed, and the LHC ran at the 13 TeV I mentioned earlier. It may be possible to push this up to 14 TeV, but it asks a lot of the magnets. We are now taking data at collision energies far higher than ever achieved before, and there is a long list of things that might show up. I'll explain what some of these are after taking a look at the least familiar member of the Standard Model which has some surprises of its own: the neutrino.

CHAPTER 8

EVERY NEUTRINO HAS ITS DAY

There are a lot of stars in the universe – more than 100 billion in our galaxy alone. This is another one of those large numbers that are hard to make sense of, so here's another way to think about it. There are around 7 billion people on Earth, and if every single person on the planet were to count one star every second, it would take only two and a half minutes to tot up every star in the Milky Way. But this is just one of the billions of galaxies out there. Carrying on at the rate of one per second, it would take the 7 billion people on Earth another 40,000 years to count every star in the visible universe. There really are a lot of stars out there.

Stars, like everything else, are made of atoms. A lot of them. There are as many atoms in a raindrop as there are stars in the universe, so counting at a rate of one per second, it would take everyone on Earth 40,000 years to count all the atoms in just one raindrop. Scale that up to the number of atoms in a glass of water, or the whole Earth, or all the stars in the universe and . . . that's a lot of electrons, up quarks and down quarks. There is an absolutely staggering number of particles in the universe.

But there is one fermion that is even more plentiful than all the quarks and electrons in every atom in the whole universe: the neutrino. This most mysterious of particles has already revealed the first cracks in the Standard Model, and may hold the key to several unanswered questions in physics. And learning more about them means leaving the LHC far behind.

Neutrinos everywhere

The neutrino was first imagined by Wolfgang Pauli in 1930 as a way to balance the energy released in beta decay, but he didn't think the 'little neutral one' could actually be measured. Some of Pauli's contemporaries took their doubts further and thought there is no way such a strange undetectable particle could even be real.

What makes neutrinos so hard to measure is that they don't carry electric charge, and so don't feel the electromagnetic force: the force that we use to interact with the world around us, and that is used by every particle detector in one way or another. In other words we, and every particle detector, are blind to the neutrino. Neutrinos only interact via the weak force, which means they barely interact at all, but because the weak force is crucial in the decays of so many other particles, neutrinos are still produced in huge numbers.

We know this happened in the first fraction of a second after the Big Bang. Particles smashing together at high energy, creating heavy tau leptons, W and Z bosons, top quarks and Higgs bosons. In the very dense early universe, some of these particles would have quickly met their antiparticle partner and turned back into a boson. After a fraction of a second, the rest will have decayed. And in any decay that involves a W boson, that W may decay into a charged lepton like a muon, and its neutrino partner. Trillions and trillions of neutrinos were produced in the Big Bang, and are now just drifting through the universe.

There is another source of neutrinos today, and it has nothing to do with the Big Bang. As the universe expanded and cooled, the surviving protons and electrons began to form the most basic atom, hydrogen. Not much else happened for millions of years, but gravity was slowly working away. Hydrogen began to gather in clouds, and into dense clumps within those clouds. As these clumps became denser and denser they would eventually ignite into a fireball: the first stars were formed, something like 100 million years after the Big Bang.

Stars are giant nuclear-fusion reactors. The force of gravity inside a star overwhelms the electromagnetic repulsion between protons, crushing them together. So close that the strong force takes over and binds them together to form a new atomic nucleus, releasing energy at the same time. If

Jupiter were around 10 times more massive it would also start fusing protons, and we would have two stars in our solar system. If we could crack the technology to achieve fusion here on Earth without these huge gravitational forces, it would provide an almost limitless, non-polluting supply of energy. There have been a number of notorious false claims and hoaxes over the years, but now major international collaborations like ITER, the International Thermonuclear Experimental Reactor, based in France, may finally be on the path to making fusion power a reality.

The first result of fusion is a new nucleus containing two positively charged protons, and two positively charged particles will repel each other. This nucleus therefore becomes more stable when one of those protons undergoes a beta decay, converting into a neutron and releasing some more energy in the form of an antielectron and electron–neutrino. The fusion continues, building up to the nuclei of helium (2 protons, 2 neutrons), then to lithium (3 protons, 5 neutrons), and so on, all the way up to iron (26 protons, 30 neutrons) – releasing more energy and neutrinos all the time. So neutrinos pour out of stars in all directions, and in huge numbers. Even though we are 50 million kilometres away from the Sun, around 10 trillion fusion neutrinos fly through every single person on Earth, every second. If a star explodes in a supernova, a rush of last-minute fusion happens, producing the nuclei of elements even heavier than iron and scattering them through the galaxy, where they eventually form planets like the Earth. Supernovae can appear as one of the brightest things in the night sky, producing a flash of light brighter than the entire galaxy the star lived in. But however much energy is contained in that flash of light – and it's a lot – around 100 times more energy is released from every supernova in a final deluge of neutrinos.

Hard to detect and ghostly they may be, but neutrinos really are everywhere. Their interactions and behaviour had a huge influence on how the universe formed and continues to evolve. As the Standard Model proved successful in describing the subatomic world, finding out more about these strange particles became more and more important.

Neutrino knowledge

Let's start with what is known about the neutrinos in the Standard Model. Neutrinos come in three 'flavours', one for each of the charged leptons, as they are the weak-force partner to the electron, muon and tau. Neutrinos don't carry electric charge or colour charge, so they only interact with the W and Z boson. We also know that they are much much lighter than any other fermion, and in the Standard Model they are required to have no mass at all. Like most things in particle physics, this is due to another symmetry: parity.

Going back to the discussion in chapter 7, something is symmetric if it is unchanged by a transformation, and the parity transformation is just reflection in a mirror. Many things look the same as their mirror image: simple shapes like a circle, and everyday objects like footballs, skyscrapers, and so on. But there are also plenty of things that don't, and there is an example right in front of us: our left and right hands are mirror images of each other. The name of mirror-image symmetry in physics is chirality, from the Greek word for 'hand'. Our hands come in two 'chiral states' – left and right – and a reflection, or parity transformation, makes one look like the other.

Given that particles are infinitely small, reflection would not seem to matter: the mirror image of an infinitely small dot is still an infinitely small dot. But there is one property of particles that hasn't come up yet, and that is spin. Even though they appear to be infinitely small dots, the fundamental particles also behave as if they are spinning. Infinitely small spinning tops. Wolfgang Pauli introduced this idea in 1924, then in 1928 Dirac discovered his equation, and this contained spin naturally (it is wrapped up in the $i\gamma.\partial$ part), showing that it is an intrinsic property of every fermion. Bosons also spin, and at double the speed of fermions – apart from the Higgs boson, which is the only non-spinning fundamental particle we know.

Spin comes with a direction. Hold out your right hand as if about to shake hands, then curl your fingers into a fist. If you imagine your thumb as the centre of a clock face, your fingers move anticlockwise around your thumb. Repeat this with your left hand, and your fingers move clockwise. All clocks are actually 'left-handed': the hands spin clockwise. We can now imagine looking at a particle: if it is moving towards us and spinning clockwise it is 'left-handed', if it is spinning anticlockwise it is 'right-handed'. These two spin directions are related by a parity transformation: if you look at a normal 'left-handed' clock in a mirror, the hands in the mirror image move anticlockwise – it becomes 'right-handed'.

For the electromagnetic force and the strong force this is not so important, as photons and gluons interact with fermions spinning in either direction. The weak force is different, as the W bosons will only interact with left-handed particles. And the Higgs boson is different again: the constant dragging through the Higgs field that gives mass to the fermions also causes them to constantly switch state, flipping between left- and right-handed. The strong and electromagnetic forces greet particles with a friendly hug, the weak force extends a tentative left-handed handshake, and the Higgs field is the gregarious uncle that picks them up and flips them around.

We don't exactly know why the weak force should distinguish between left- and right-handed particles, though probably a reasonable answer is simply because it can. Chiral symmetry is possible in the universe, and almost every possible symmetry we can imagine is realised somewhere; chiral symmetry appears in the weak force. We can describe this very well in the Standard Model, but we don't yet have a deeper explanation for why it happens there.

And for the neutrinos, it leads to an interesting situation. They only interact with the weak force, which only sees left-handed particles. In order to interact with the Higgs field and pick up mass, a particle must have both hands, but the neutrino is so light it doesn't seem to do this. An explanation for all of this might be that there is just no right-handed neutrino – and this is how the Standard Model was built: neutrinos are massless, and only spin in one direction.

Finding neutrinos

It's worth saying again: neutrinos are really very hard to measure. They seem to have no mass, which means they travel at the speed of light, and will interact so rarely they will fly through any material without touching it. But in order to detect a particle and learn about its properties, it has to interact, to hit something. It is possible to work out how frequently neutrinos interact (not very), and so how large a detector would have to be in order to have a good chance of a neutrino hitting it: more than a light-year long. That is, if we fired a neutrino into our detector, this detector would have to extend a quarter of the way to the nearest star, Alpha Centauri, to have even a 50% chance of detecting it.

This is just yet another way of saying that it is extremely difficult to detect neutrinos. When they are produced at the LHC in the decay of a heavier particle, they fly out of the collision unnoticed. This still gives a characteristic signature though, as collisions are usually balanced: conservation of momentum tells us that when some particles leave in one direction, there must be one or more particles with equal momentum leaving in the opposite direction. The neutrino appears as 'missing energy': because we don't measure the neutrino, a collision will appear unbalanced. By measuring everything else, we can calculate fairly precisely how much energy the neutrino had, which direction it was travelling, and then how it might have been produced.

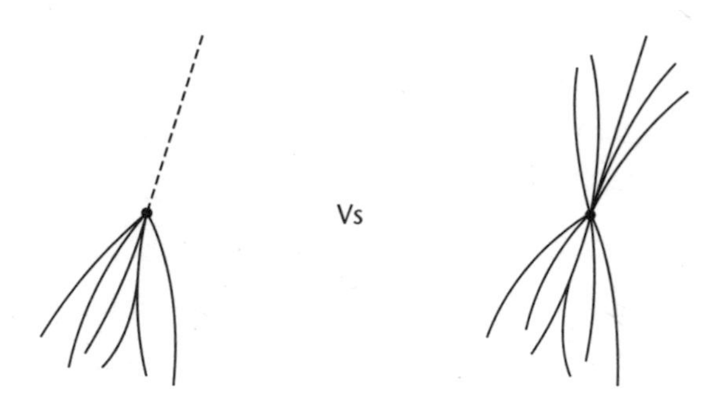

Particles leaving a collision that produced an undetected neutrino (represented by the dashed line), compared to one without a neutrino.

But missing energy doesn't tell us much about the neutrino itself: which of the three possible types (electron, muon or tau) it might be, or how it might behave. To really study neutrinos, we need to catch them in action, and this is definitely a numbers game. Rather than building a detector light-years long to try to measure a single neutrino, we rely on the fact that there are trillions of them all around us all the time, and every once in a while just one of the trillions will do something we can measure.

So in order to learn anything about neutrinos we need a lot more of them. There are four main factories that churn them out in sufficient numbers, giving us four kinds of neutrino to study:

1) Solar neutrinos. The most plentiful source, these are electron–neutrinos produced by the nuclear-fusion processes happening in the heart of the sun.

2) Reactor neutrinos. Nuclear power stations run on the energy produced in nuclear fission, the breakup of large unstable atomic nuclei. They mainly feature the opposite process to the fusion in the sun, producing electron–antineutrinos. The very first detection of neutrinos was by Cowan and Reines's 1956 experiment, located next to a reactor in South Carolina. They were able to detect around 3 neutrinos per hour (well over a thousand million million million were flying straight through the detector in that time).

3) Atmospheric neutrinos. When a high-energy cosmic ray hits the Earth's atmosphere, a shower of particles is created. These reactions happen at a much higher energy than the nuclear processes in the Sun and in nuclear reactors, and the heavy muon is produced plentifully in the weak decays of charged pions and kaons. The result is mainly muon–neutrinos, and some electron–neutrinos, for example:

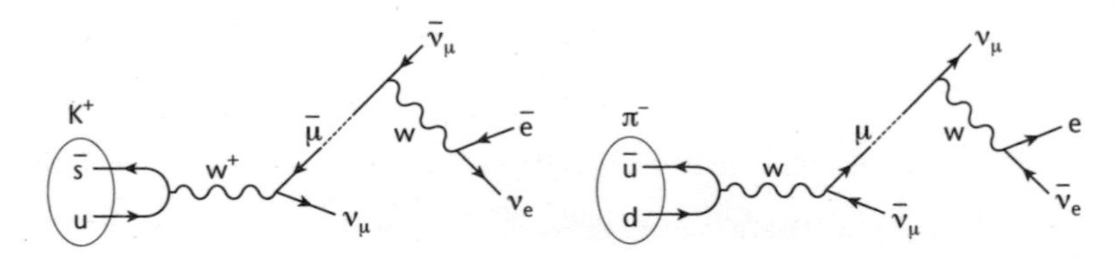

The decay of a kaon and pion. The muon can travel several kilometres before also decaying.

4) Accelerator neutrinos. Taking a beam of accelerated protons, like the beam from the LHC, and firing it into a solid target mimics a cosmic-ray collision. Pions and kaons are produced, which decay to muons – and lots of neutrinos. The advantage is that it is possible to produce a concentrated beam, with control over both the direction and energy. It is also possible to produce tau–neutrinos by firing a higher-energy proton beam at the target, making heavier particles that can decay to taus (and their accompanying neutrino). Tau–neutrinos were finally observed in the DoNuT accelerator neutrino experiment at Fermilab in the USA in 2000, the last Standard Model fermion to be discovered.

Everyone's favourite neutrino story also involved accelerator neutrinos when in 2011, for a short time, it seemed they might be breaking one of the most fundamental rules in the universe by travelling faster than the speed of light – something that has been thought impossible ever since Einstein's Special Theory of Relativity in 1905. A particle accelerator at CERN was used to produce a beam of neutrinos, which was aimed directly beneath the Alps and south around 730 km to an underground lab in Italy, where the OPERA experiment (Oscillation Project with Emulsion-tRacking Apparatus) would detect them. Travelling at the speed of light, this journey should take them around 2 millionths of a second, but they were arriving around 60 billionths of a second early. This would have been truly revolutionary if correct, challenging one of the foundations of modern physics.

I have no connection with OPERA, but other anomalous

results have come up on experiments I have been involved with. When something unusual like this occurs, it will be noticed long before the experimenters make the result public. The first assumption is that some source of systematic uncertainty was missed in the analysis – a mistake in the calibration, or maybe the modelling; it could be anywhere, so numerous thorough checks are made. But after exhausting all possible sources, what do you do? It goes against the scientific method to discard a result you don't like, and besides, making the result public opens it up for more scrutiny, and more ideas for where the problem may lie. And this is what OPERA did – when I saw the results presented, it was not as a claim of revolutionary new physics, but as a strange result that must have a more mundane explanation. And in the end, that's what it was: synchronising and calibrating equipment to the required level of precision over that huge distance is immensely challenging, and within a year a problem had been found – it turned out to be a small timing error. Neutrinos may be odd, but they are not that odd!

The problems with neutrinos

Having picked a source, the next thing to work out is how to measure the neutrinos. And to do that, they have to interact with something.

The problem then is the low rate of detectable events – in order to get a satisfactory number of interactions, detectors have to be huge. One of the most famous early neutrino detectors, the 'Ray Davis Experiment', named after its main scientist, used a giant cylindrical tank 6 metres across and 15 metres long holding around 800 tons of dry-cleaning fluid – which contains a lot of chlorine. The idea was to detect solar electron–neutrinos undergoing a W interaction. The W emitted by a neutrino would be absorbed by a neutron in the nucleus of a chlorine atom, converting it to a proton. This turned chlorine atoms, which contain 17 protons, into argon, which contain 18. Argon is a chemically inert gas, so simply floated to the top of the tank.

Even with this huge tank, with trillions of neutrinos flying through it every second, fewer than two argon atoms would be produced on an average day. Because this number is so small, the experiment had to be shielded from any background radiation and cosmic rays, which might produce false signals, so it was set up 1.5 km underground, down an old gold mine in Homestake, South Dakota. Every few weeks the tiny amount of argon was collected, making it possible to calculate the number of neutrino interactions that had happened in that time.

The experiment ran from 1970 to 1994, and while a giant tank of dry-cleaning fluid down a mine is a very different type of particle physics experiment from the LHC, it was able to do something the LHC could not: provide precise measurements of electron–neutrino interaction rates. So precise that it turned up an unexpected result: given everything we know about the Sun, there should have been a lot more neutrinos detected. About a third of them were missing.

This became known as the solar neutrino problem, and after many checks of the experiment itself, it seemed that there was a big problem in our understanding either of how the Sun works, or of how neutrinos behave.

It took another experiment to solve this problem, this time down an old nickel mine in Sudbury, Canada: the Sudbury Neutrino Observatory (SNO), which ran from 1999 to 2006 and consisted of 1,000 tons of purified 'heavy water' (water, H_2O, containing a higher fraction of deuterium, a hydrogen isotope with a proton and a neutron in the nucleus instead of just a proton). SNO had two main advantages: it was able to detect individual neutrino interactions in real time instead of accumulating results over weeks, and it was sensitive to neutrino interactions involving either a W or Z. In the W interactions, an electron–neutrino converts into an electron; the W is then absorbed by the detector. The Z interactions allow neutrinos of any flavour to knock electrons out of a nearby atom, or to break a deuterium nucleus into a proton and neutron. These Z interactions lead to two signals: gamma rays, produced when the neuron was captured by another nucleus; and individual electrons.

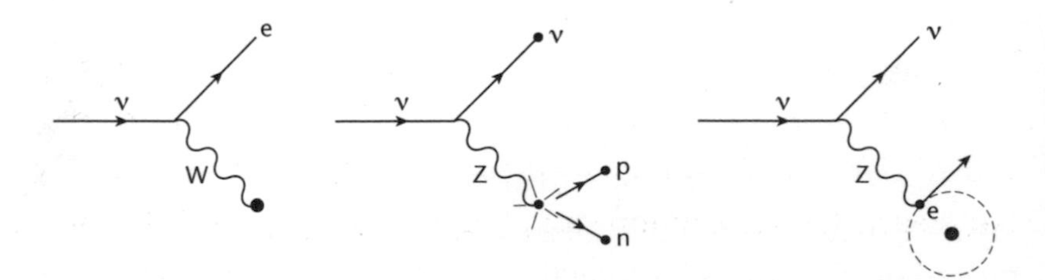

The three different neutrino interactions with the SNO detector.

Gamma rays are easy to pick up as they fly out of the detector. Single electrons don't make it that far, and picking them out in a huge tank of water is impossible, so they were measured using Cherenkov radiation. Named for its discoverer, (Pavel) Cherenkov radiation is the electromagnetic equivalent of a bow wave in water, or a sonic boom in the air. As light travels through any material, it travels a little slower than in empty space. If you look at a straw in a glass of water, the straw appears to bend where it enters the liquid – an optical illusion caused by light changing direction as it slows down in the water relative to the air. High-energy particles, which travel very close to the speed of light in a vacuum (the real speed limit in the universe), can be travelling faster than light in a material like the liquid detector at SNO. When this happens, Cherenkov radiation, typically a blue light, is produced, just as an aeroplane creates a sonic boom when travelling faster than sound in air. Light detectors (photo-multiplier tubes, or PMTs) pick up both the gamma rays and the Cherenkov radiation, thereby detecting the neutrinos.

Disentangling the W interactions (only sensitive to electron–neutrinos) from the Z interactions (sensitive to neutrinos of any flavour, electron, muon or tau–neutrino), SNO found the problem: the Sun should only be emitting electron–neutrinos, but when they reached the Earth there were around ⅔ electron–neutrinos and ⅓ the other flavours (muon– and tau–neutrinos). The muon– and tau–neutrinos were then not picked up through W interactions, as they lacked the energy to turn into their much more heavily charged partner. This result meant that neutrinos were apparently changing flavour while travelling, and without interacting with anything else. No other particle in the Standard Model can do this, and there is no interaction, no Feynman Diagram you can draw for the electromagnetic, strong or weak force that allows it to happen. Something new was going on here.

In the meantime, another giant water detector in a mine in Japan had found another problem with atmospheric neutrinos – that is, the muon–neutrinos produced by cosmic rays. Because atmospheric neutrinos tend to be higher-energy and less abundant than solar neutrinos, a much larger tank was needed. The 3,000-ton Kamiokande detector was completed in 1987, and was followed by the huge Super-Kamiokande in 1996, a cylinder roughly 40 metres across and 40 metres high, holding 50,000 tons of purified water, 50 times the mass of the SNO. There are amazing photographs available online from the construction and commissioning of Super-Kamiokande showing a team of scientists in a small inflatable dinghy cleaning the PMTs around the edge of the tank as it is slowly filled.

But even with this huge tank, interactions are still incredibly rare. Most neutrinos just fly straight through the Earth and out to space without doing anything. Also, because cosmic rays are hitting the Earth from all directions, the atmospheric neutrinos were coming from all directions: straight down if the cosmic ray hit the atmosphere above Japan, straight up if it hit the other side of the world, and every angle in between. By comparing neutrinos from different directions, another problem was found: the rate of neutrinos going down was roughly what was expected, but the rate coming up was too low. This is the atmospheric neutrino problem, and again further experiments were able to show that the neutrinos were still there, but that they had just changed flavour from muon–neutrinos into something else.

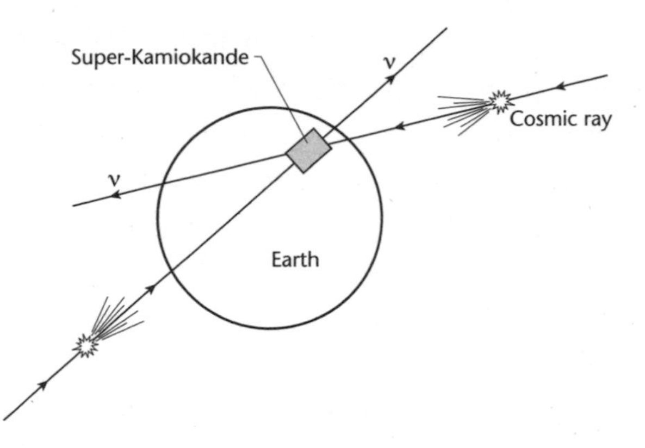

Super-Kamiokande can detect neutrinos from cosmic rays in all directions (not to scale!).

The solar and atmospheric neutrino problems tell us that neutrinos really do change flavour mid-flight, something forbidden by the Standard Model. This discovery has proved so important in our understanding of the ghostly neutrino that the 2015 Nobel Prize for Physics was awarded to Arthur B. McDonald from the SNO and Takaaki Kajita from the Super-Kamiokande experiment. It is sometimes called the first sign of physics beyond the Standard Model, because it tells us that neutrinos must have mass.

Oscillating neutrinos

The solution to the flavour-changing neutrinos is a fairly subtle quantum effect that requires three ingredients we have met earlier in the book. The first is mixing, which came up in chapter 5 for the quarks: the quarks are arranged in pairs, but these don't perfectly correspond to the weak-charge pairs. This means that a W boson can mix up quarks from different pairs, converting a charm quark into a down, for example. At the time, I said this doesn't happen for leptons. And if neutrinos are massless, this is true. But if neutrinos do have mass, mixing is possible.

The second ingredient is related to how particles move, and as explained in chapter 6, this involves some imaginary pedometers with one hand that winds as the particle travels. There is one part I did not explain before, but that becomes important here: the speed a pedometer winds actually depends on the particle's mass. For heavy particles, it winds faster than for a lighter particle travelling the same distance.

The final ingredient goes all the way back to chapter 2: when we are not measuring a particle, it will do everything, all at once. For a neutrino, we can consider a weak interaction to be a 'measurement', as we can detect these interactions. So, in between emitting W and Z bosons, a neutrino is doing everything, all at once.

Now to combine these ingredients, and solve the missing

neutrino problem. Let's consider a solar neutrino, produced when a proton converts into a neutron, emitting a W. The W decays into an antielectron and a neutrino: the first measurement. Now, because leptons can actually mix, this neutrino does not actually have to be an electron–neutrino; it might be a muon–neutrino or a tau–neutrino. And because we are not measuring the neutrino directly, we have to consider all of these possibilities, all at the same time.

The neutrino now flies towards the Earth. As it does so, the little imaginary pedometers are winding. But because we have to consider all three possible neutrinos, all at the same time, we can picture this as one pedometer with three hands. The three possible neutrinos have different masses, so the three hands move at different speeds.

You can think of this like three cars racing around a circular track. Let's call the cars 'e', 'mu' and 'tau', and at the start of the race, the cars are lined up one behind the other. But they travel at slightly different speeds, with the 'tau' car slowly pulling ahead of the others. Later in the race the cars can get mixed up, as the 'tau' car can catch the others and for a while lag behind them on the circuit, before eventually overtaking and going ahead again.

Now, at different points in the race we may ask: Which is the next car to cross the start line? At the start of the race, it will be the 'e' car. Later in the race, the 'tau' car will have pulled ahead. Later still, the 'tau' car will have caught the others, appearing to be behind them on the circuit, and the 'e' or 'mu' car are more likely to cross the line next. In other words, the answer changes as the cars travel further, completing more and more laps of the circuit, and changing positions relative to each other.

In this analogy, asking which car will be next to cross the start line is equivalent to asking how a neutrino will behave in the next measurement: emitting a W and turning into an electron, muon or tau. At the start of the race, we have an electron–neutrino: the 'e' car is ahead of the others, so crosses the line first. But because the cars travel at different speeds, then the cars on the circuit change ordering; neutrinos change flavour after travelling some distance. What started out as an electron–neutrino may become a muon–neutrino. And this is what happens to the solar and atmospheric neutrinos – and to all neutrinos: they 'oscillate' from one flavour to another as they travel.

In order to make all this work, several new pieces of information have to be added to the Standard Model. First, there are the three different neutrino masses, which correspond to the three different speeds of the cars. If the three masses were all the same (or indeed all zero, as was first thought), the cars would travel at the same speed round the track and remain perfectly synchronised – an electron–neutrino would remain an electron–neutrino, because the 'e' car would start in the lead and always cross the line first.

Then there is the amount of mixing, or how likely a W boson is to mix up an electron with a muon–neutrino, and so on. There are four numbers that control the various amount of mixing between the different kinds of neutrino, and the information is contained in something called the PMNS matrix (after Bruno Pontecorvo, Ziro Maki, Masami Nakagawa and Shoichi Sakata). If all four numbers were zero, then there would be no mixing – equivalent to there being just one car on the circuit: if we start with an electron–neutrino, we will always have an electron–neutrino. If the numbers are small, then the other cars spend most of the race in the pits and only get on the track occasionally, so the neutrinos rarely change flavour. If the numbers are large, all the cars are on the track all the time.

While the theory tells us that mixing can happen, it doesn't tell us anything about the numbers that control it. They are actually 4 of the 26 free parameters in the Standard Model that I'll come back to in the next chapter. There is now a wide range of neutrino experiments, using reactor and accelerator neutrinos as well as solar and atmospheric, measuring these numbers, which will ultimately tell us exactly what role neutrino oscillations may have played in shaping the universe. But there are also some more fundamental questions about the mysterious neutrino.

Anti-antineutrinos

Having dealt with the facts we currently have about neutrinos, it's time to enter the realm of speculation. First up, the fact that a neutrino does not carry electric charge raises a strange possibility: maybe it is its own antiparticle.

For every fermion in the Standard Model, the particle and antiparticle have opposite charges. The positively charged positron is the antiparticle to the negatively charged electron; the negatively charged antiup quark is the antiparticle to the positively charged up quark, and so on. But perhaps the neutral neutrino and the neutral antineutrino are actually the same thing? A particle that is also its own antiparticle is known as a Majorana fermion, a variety of the particle named after Enrico Fermi that is now also named after Ettore Majorana, who suggested the possibility in 1937; to make the distinction, all other fermions are technically Dirac fermions. It's an interesting idea, but given how rare neutrino interactions are, it is almost impossible to tell if they are Majorana particles or not. The best chance we have of finding out is something called neutrino-less double beta decay.

Double beta decay does occur naturally in some radioactive isotopes, where a nucleus can become more stable when two neutrons convert to protons simultaneously. This is the same beta decay we have seen before: quarks emitting W bosons, which decay into an electron and an antineutrino. But if the neutrino is its own antiparticle, something else could happen: the neutrino from the first beta decay could play the role of an incoming antineutrino for the second beta decay. In this case, two electrons and no neutrinos would be produced. This is neutrino-less double beta decay, only possible if the neutrino is its own antiparticle.

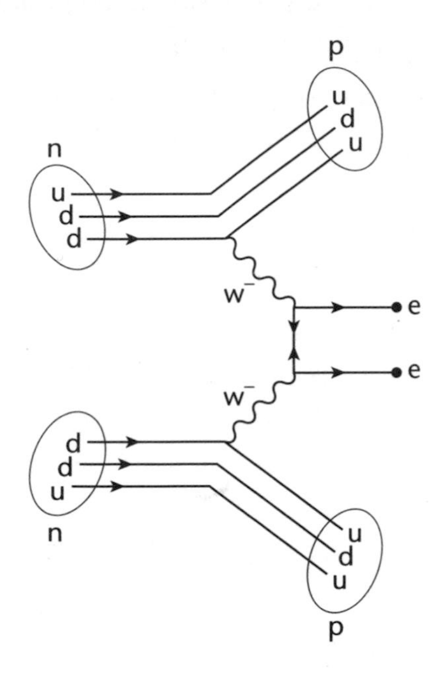

A Feynman Diagram for neutrino-less double beta decay.

It's not easy to tell the difference between the normal and the neutrino-less varieties of double beta decay, as we have such an incredibly low probability to measure any neutrinos if they are there. Instead, the two electrons hold the key: in the neutrino-less case, they carry away all the energy of the interaction; in the case with neutrinos, the same energy is being shared between four particles, the two electrons and two neutrinos. Measure the two electrons precisely, and it should be possible to spot the neutrino-less case.

This requires incredibly precise electron measurement and

an incredibly clean environment. Neutrino-less double beta decay, if it occurs at all, will be extremely rare, and any random beta decays from natural sources of radiation could provide false positives that spoil the precision of the experiment. The amount of natural radiation from a bunch of bananas would be too much, and the scientists must stay well clear – the average person is 350 times more radioactive than a banana!

The state of the art for detecting double beta decay is SuperNEMO – the bigger-budget sequel to NEMO, the Neutrino Ettore Majorana Observatory, which ran from 2003 to 2011. This is a different kind of experiment: a large sample of radioactive material surrounded by high-precision electron detectors. SuperNEMO is currently under construction in a mine under the French Alps, and in the next few years may be able to determine if neutrinos really are their own antiparticle – making them even stranger.

The right hand

The original formulation of the Standard Model stated that neutrinos come only as left-handed particles. For any particle to pick up mass from the Higgs field, it must come in both left and right-handed versions, so the neutrino also had to be massless. Oscillations showed that neutrinos do have mass, so does this mean there must also be a right-handed neutrino – and if so, where is it? Or do neutrinos get their mass from something other than the Higgs field?

The right-handed neutrino would be a very strange particle. It would not interact with any of the forces in the Standard Model: it doesn't carry electric or colour charge, and being a right-handed particle, it doesn't feel the weak force. For this reason, a right-handed neutrino (and there must be three of them, one for each of the three left-handed neutrinos) is called a sterile neutrino.

However hard it is to detect the regular, left-handed neutrino, detecting the right-handed one would be much worse. Impossible, in fact. But it may have an influence on neutrino mixing: if a neutrino oscillates into a sterile right-handed state, this can change the probabilities of detecting the three flavours of left-handed neutrino. Experiments are working to pin down these oscillations to look for any sign of this sterile neutrino effect.

If neutrinos do pick up mass from the Higgs field, the final puzzle is why they are so very much lighter than every other particle – over a million times lighter than the next-lightest particle, the electron. There is a possible explanation for this, but first a word on antineutrinos. Back in chapter 2, I introduced antimatter by rotating a Feynman Diagram. A particle travelling into a vertex becomes an antiparticle travelling out, and antiparticles are equivalent to particles travelling backwards in time. Now, if you take some video of a clock and play it backwards, then the clock hands will move backwards: anticlockwise. A left-handed clock moving forwards in time becomes a right-handed clock moving backwards in time. The same is true for particles: the antimatter partner of a left-handed neutrino is a right-handed antineutrino. If there is also a right-handed neutrino, it will have a left-handed antineutrino partner.

This is getting confusing, especially because we don't normally distinguish between left and right-handed versions of particles. But neutrinos are different, and if neutrinos are Majorana particles, then things get both a bit simpler and a bit more complicated. Majorana particles are their own antiparticle, but if there is a right-handed neutrino then we still have two states here: the left-handed neutrino, which is the same particle as the right-handed antineutrino; and the right-handed neutrino, which is the same particle as the left-handed antineutrino.

So, if both of these states do exist, something called the see-saw mechanism can happen. Instead of the left- and right-handed neutrinos combining and ending up with a mass similar to the other Standard Model particles, the balance can tip. The left-handed neutrino can become very light (which is what we see, and something we would like to explain), and the right-handed extremely heavy – around a billion times heavier than any particle we know, and certainly far beyond the reach of any current experiment (which again matches what we see, or in this case don't see – we have never measured a right-handed neutrino). This is a neat explanation, but for it to be true we first need to determine if neutrinos are Majorana particles.

Neutrinos hit the big time

The extraordinary difficulty in measuring neutrinos has led to some extraordinary ideas for neutrino detectors. If having a large water detector is the key to detecting these strange particles, we can just use the sea. This is what the ANTARES telescope does, sitting around 2.5 km underwater off the south coast of France. It consists of a series of light detectors, PMTs, suspended on vertical chains, looking for Cherenkov radiation from any neutrino interactions in a huge volume of water. There are plans to build the much larger KM3NeT (cubic kilometre Neutrino Telescope) in the future, and while there are difficulties dealing with a detector on the bottom of the seabed, not least of which are all the other creatures sharing the water, looking in such a huge volume brings sensitivity to the hunt for neutrinos at very high energies.

Of course, there is not much living under the ice in Antarctica, so there are also neutrino experiments there. The IceCube experiment consists of chains of PMTs dug deep into the ice at the South Pole, spread out to cover a cubic kilometre and able to pick up any Cherenkov radiation from neutrino interactions in the ice. Going one step further, the ANITA experiment suspends detectors from a helium balloon around 35 km above the South Pole, where it can detect any neutrino interaction in the ice (producing radio signals this time) over a huge area – effectively turning most of the southern ice cap into a giant neutrino trap.

The advantage of these huge detectors is that they are sensitive to incredibly high-energy neutrinos, produced in the most violent areas of the universe: supernovae, active galactic nuclei, black holes and many other interesting phenomena. While neutrino astronomy is not going to replace the more traditional photon astronomy any time soon, it does offer a different perspective. Some photons cannot reach the Earth, due to gas clouds or other things blocking the way; neutrinos, of course, mostly just fly straight through these obstacles, giving us a direct line of sight to some of the most spectacular events in the cosmos. There has only been one supernova in or close to the Milky Way since neutrino detectors have been running: SN 1987A. Seen on Earth in 1987, it occurred 160,000 light-years away (and therefore 160,000 years ago). Kamiokande detected 11 neutrinos in around 13 seconds from this supernova: it would normally have taken around a week to see this many from the Sun, which is a mere 8 light-minutes away. This highlights again the staggering number of neutrinos produced in a supernova: estimates for SN 1987A put it at something like 10^{58}. The neutrinos also arrived a couple of hours before the light from the supernova – not because neutrinos travel faster than the speed of light, but because the photons get scattered along the way, taking longer to escape the explosion. The large neutrino experiments around the world today are not only probing the Standard Model in new ways, they are also ready to provide an early warning for astronomers looking for the next supernova!

A different kind of particle physics

Neutrinos are fascinating. They really are the ghosts of the Standard Model, and for so long were out of reach. Getting to know them requires a completely different approach from that of the LHC: giant tanks of water hidden down mines; beams of particles fired through kilometres of solid rock; observing the nuclear reactions happening in the centre of the Sun; detecting gamma rays and Cherenkov radiation. Instead of sifting through millions of collisions every second, as at the LHC, the detectors have to be so quiet that a handful of events can be seen.

But neutrinos are now revealing new things about the universe, proving once again that nature is stranger than we expect. And neutrino detectors also form the template for the experiments searching for another particle that interacts very rarely: dark matter. The nature of this dark matter is one of the big open questions in physics today, and it is to those questions that we turn next.

CHAPTER 9

INTO THE DARK

Particle physics is now entering uncharted territory. Since it was constructed in the 1960s and 70s, the Standard Model has pointed the way: more quarks and leptons to be discovered, the W and Z, then finally the Higgs boson. Neutrino masses were a surprise, but have been added to the story. And that story is now pretty much complete: 12 particles, 3 forces, the Higgs mechanism, and an amazingly successful description of the subatomic world.

In this chapter, I will place the Standard Model in context. Moving on from the amazing successes described earlier in this book, it is time to focus on the open problems, the reasons we know the Standard Model is incomplete. To see what happens when we take the Standard Model's predictions, scale them up and compare what we find with the world around us. And they are big problems.

Where is all the antimatter?

It's something that we take for granted: the world is made of matter. Dirac first predicted antimatter as one of the consequences of his 1928 equation, but this led to the obvious question: how can this strange stuff exist if we have never seen it? Now antimatter is an essential part of the Standard Model, but matter and antimatter are always made in equal amounts. A high-energy photon can turn into an electron and a positron. A high-energy gluon can turn into a quark and an antiquark. One matter particle and one antimatter particle. Today, everything is made of just matter, no antimatter. So the question about antimatter has been turned on its head: where did it all go? How did we manage to lose half the stuff in the universe?

There is a possible explanation, but it requires matter and antimatter to behave slightly differently. Just after the Big Bang, the universe was filled equally with matter and antimatter. High-energy photons, gluons, W bosons and Z bosons and Higgs bosons were being created, and were decaying into particle–antiparticle pairs. But if there was a tiny difference in the way that matter and antimatter behaved, then this 50% matter and 50% antimatter balance may have tipped slightly towards matter, maybe 50.001%–49.999%. Most matter and antimatter particles met up and annihilated in the hot dense early universe, turning back into bosons. But because of this very slight imbalance, a small amount of matter was left over. This 'tiny' remainder now makes up all of the universe we can see.

To make this picture fit, there must be a small difference between matter and antimatter. In the language of the Standard Model, there must be some interactions, a Feynman Diagram we can draw that will separate the two. But this is the problem: there isn't.

I first introduced antimatter back in chapter 2, by simply rotating a vertex in a Feynman Diagram. Doing this took an electron moving forwards in time and turned it into an electron moving backwards in time. This antielectron has the opposite charge to an electron, but is fundamentally still an electron – a car is still fundamentally a car, no matter which way it drives up the street.

Particles also spin, and come in left- and right-handed versions – though only the weak interaction can really tell the difference (the weak interaction is the source of so many difficulties in the Standard Model!). As I mentioned in the last chapter, changing the direction a particle moves in time also performs a parity transformation: the W bosons only interact with left-handed particles, and right-handed antiparticles.

So if we look at every interaction, every Feynman Diagram possible in the Standard Model, and swap particles for antiparticles, this is equivalent to swapping their charge and parity, or CP. And under this change, every interaction we can calculate will remain unchanged – this is known as CP-symmetry. According to CP-symmetry, matter and antimatter really behave in identical ways, and if CP-symmetry is true, the universe should be half matter, half antimatter.

If we are to explain why our universe only contains matter, to explain why the antimatter vanished, this CP-symmetry must be broken somehow: a process known as CP violation. Trying to explain the absence of antimatter in the universe has mainly involved searching for CP violation in particle interactions.

Although I said the Standard Model does not violate the CP symmetry, there is actually one place it can happen: mixing, which we met in chapter 4 for quarks, and chapter 8 for neutrinos. Mixing is a slight misalignment between the pairs of fermions in the Standard Model (electron and electron–neutrino, up and down quark, for example) and the weak-charge pairs. This means that W bosons can mix between quark pairs, allowing all heavy quarks to decay, and between lepton pairs, allowing neutrinos to change flavour when flying. And it turns out that this mixing can occur slightly differently for matter and antimatter.

Quark mixing has been studied extensively, and CP violation has been confirmed to exist. Matter and antimatter do behave differently. An example is in the decays of the kaon, which can decay either to a positive pion, negative electron and electron–antineutrino, or to a negative pion, positive antielectron and electron–neutrino. These two different decays should happen at the same rate, but the decay producing an electron happens around 0.3% more often than the decay producing a positron. This is the kind of effect we are looking for, but like every sign of CP violation with quarks, it only affects a few particles, and is far too small to explain the matter-dominated universe we live in.

If the quarks are not the answer to this problem, perhaps it is over to the neutrinos. They may break the CP-symmetry to a much larger degree, but because neutrinos are so much harder to measure, we don't yet know. Finding out is one of the top priorities for particle physics, and the target for several upcoming neutrino experiments. And if the neutrinos also don't break the CP-symmetry by a large amount, then we will need a new explanation for the missing antimatter.

There is one more curious aside in the CP-violation story. In the Standard Model, it is only the weak force that treats matter and antimatter differently, but it doesn't have to be that way: the strong force could do so as well. And if it did, this could be the extra effect that pushed our universe into being completely dominated by matter. But the strong force seems to respect the CP symmetry perfectly, and at the moment we have no idea why. This is another open question in physics, known as the strong CP problem.

CP violation is not the only possible answer to the missing antimatter. Some theories suggest that antimatter responds to gravity in a different way: matter and antimatter may repel each other. There may be no antimatter in our part of the universe because the matter here has pushed it all away! However, it is extremely difficult to produce enough antimatter and measure how it responds to gravity, so while this is a strange idea, so far it is one that we cannot rule out.

The Standard Model: Why?

In many ways, the Standard Model is the model of efficiency. It describes the world using just 12 fermions – and really, only three of those are important: the electron, the up quark and the down quark, which combine to form atoms. But the Standard Model tells us the 'how' but not the 'why': how these particles behave, but not why these particular 12 exist.

There are many ways to combine the Dirac Equation with gauge symmetries and the Higgs mechanism, and it's not clear why nature chose the particular combination in the Standard Model. Why there are three forces – the electromagnetic, weak and strong. Or why the weak force distinguishes between left- and right-handed particles. Why quarks carry fractional electron charge, the electron a whole unit, and the neutrino none at all. Why quarks carry colour charge, but leptons don't. Why there are three generations of matter, when one would seem to be enough. And we also don't yet know if right-handed neutrinos exist, or if neutrinos are Majorana particles (particles that are also antiparticles).

In other words, the Standard Model is empirical: it was built to match what we see in experiments. We understand the individual elements, but not why they are put together in this form. There must be something underlying all of this, a more fundamental description of nature. We just don't yet know what this is.

Even if we just accept the structure of the Standard Model

for now, it still contains 26 'free parameters', 26 numbers that we cannot explain, we just have to go and measure. I've mentioned them all at different parts of this book, so here they all are in one place:

- the strengths ('coupling constants') of the strong, weak and electromagnetic forces.
- the strength of the Higgs field (the energy stored in empty space, or 'vacuum expectation value').
- the mass of the Higgs boson.
- the masses of the 6 quarks and 6 leptons, which can also be interpreted as how strongly these particles interact with the Higgs field.
- three numbers controlling quark mixing, and a fourth number controlling quark CP-violation.
- another four numbers for neutrino mixing and CP-violation.
- the number controlling CP-violation in the strong force, though it seems to be exactly zero.

Once we have measured these numbers, we can plug them into the equations of the Standard Model and calculate amazingly precise predictions (including the masses of the W and Z bosons). But it's more than a little unsatisfying – 26 reasons to think there must be a deeper explanation for all of this.

To add to this 26, there are also other fundamental numbers that reach beyond the Standard Model, like the speed of light. Any theory with so many loose ends, so many things left unexplained, surely cannot be complete. Again, if we have a deeper understanding of nature, a theory that explains the structure of the Standard Model, this same theory may also explain some of these numbers.

The Standard Model: tangled up in loops

There is also something rotten right at the core of the Standard Model. This was first found in quantum electrodynamics – the most precise theory in the history of science, but in the 1930s it was the theory that couldn't predict anything. Calculations were going wrong. It seemed as though the initial promise of these new ideas was collapsing under a lot of nonsense maths. There was a problem in the loops.

Loops can happen in the slightly more complicated Feynman Diagrams (as we saw in chapter 2), diagrams that contain a boson temporarily turning into a particle and antiparticle and back, or a fermion emitting a boson then reabsorbing it almost immediately. The particles inside loops are virtual particles: trapped inside a Feynman Diagram, they are things that we cannot measure directly. The uncertainty principle allows these virtual particles to bend the rules, to borrow energy, with one condition: the more energy is borrowed, the shorter the time it can be borrowed for. And these loops have a tendency to blow up, to borrow more and more energy by becoming smaller and smaller, eventually borrowing an infinite amount of energy. When a calculation returns an infinite answer, this usually means something has gone badly wrong. And this was the state of play in the 1930s: QED was broken.

A whole branch of quantum mechanics known as renormalisation had to be developed to solve the problems with these infinities. Techniques to sweep them under the carpet and return a 'normal' answer. Renormalisation essentially says that the loops cannot really run away to infinitely high energies, because at some point a different type of interaction must take over. There are plenty of ideas for what that might be (like Super-Symmetry or string theory), but the point is that it's just not really relevant when calculating things like beta decay, or collisions at the LHC. It's like going for a drive and worrying about the behaviour of every single person driving every other car on every road in the world – yes, they're all out there, and who knows which roads we'll drive on in the future, but it's a level of detail that is not currently important, and actually makes driving so complicated it becomes impossible. In order to drive anywhere, we impose a cut-off: we just need to think about the car we are in and the others we can see, and not worry about the rest.

The same thing happens in renormalisation: a cut-off is imposed, an emergency brake to stop the loops running away to infinity. We don't really have to worry about what is happening at extremely high energy right now, we just need to say that it is something other than the Standard Model. The maths of QED breaks because we are trying to use it where it is not valid, so let's just stop it before reaching that point. Paul Dirac was never happy with this approach, feeling that this artificial cut-off spoiled the beauty of his equation. Even Richard Feynman, who pioneered many of these techniques, called renormalisation 'hocus-pocus'. But it is essential – without it we cannot calculate anything in the Standard Model.

Renormalisation really tells us that the Standard Model is just an approximation. The predictions break at extremely high energies because it simply doesn't work there. There must be a deeper explanation behind all of this. And at some point in the future, maybe at the next-generation particle accelerator, maybe far far in the future, we will discover this new thing. The Standard Model will live on as the low-energy approximation to that theory – in the same way that Newton's Laws live on as the low-energy approximation to Relativity.

Dark matter

If the free parameters and renormalisation are circumstantial evidence for a deeper theory, there is also a smoking gun. The best evidence right now is not coming from the subatomic world, but from some of the largest things in the universe: galaxies.

Galaxies are beautiful. Giant rotating spirals, discs and balls of stars, held together by gravity. We sit around two-thirds of the way out from the centre of our own galaxy, the Milky Way, and we are zooming around the centre at around 800,000 km per hour. It still takes around 200 million years to make a complete circuit – go back one rotation and dinosaurs were roaming the Earth. But this is also a problem: the Milky Way, like almost every other galaxy, is rotating too quickly. These beautiful spirals are spinning so quickly that they should be flying apart.

This problem was first noticed in the 1930s, but the work of Vera Rubin in the 1960s made everyone take notice. And it goes like this: by looking at how bright a galaxy is, it is possible to estimate the number of stars it contains (typically hundreds of billions), and hence how much mass. From the amount of mass and the size of the galaxy, it is then possible to work out the gravitational force. The stronger the pull of gravity, the faster the stars move. The same idea holds for all things in orbit: the International Space Station orbits about 400 km above the Earth and is travelling at about 7.6 km/s (over 17,000 mph); the Moon is almost 400,000 km away, and travels at a mere 1 km/s (over 2,200 mph). Drawing the expected speed for stars at different distances from the galactic centre gives a 'rotation curve'.

Now, as well as calculating how fast galaxies should be rotating, it is also possible to actually measure it, and this is where the problem lies: the measured speed is always higher than the expected, especially far from the centre of the galaxy.

One possibility is that the calculations are simply wrong, because the model of gravity being used (Einstein's General Relativity) is not the full story. There have been attempts to solve the problem this way, building a model simply by modifying the equations of gravity until they describe the data. MoND (Modified Newtonian Dynamics) is a theory that has been around since the 1980s, suggesting that gravity works slightly differently when considering the very large scales of galaxies, and it has been able to successfully describe the rotation curves.

Another explanation is simply that there is a lot more stuff in galaxies, but it is stuff that we can't see – dark matter, an idea that has been around since the 1920s. Because we can't see it, this dark matter is not counted when calculating the gravitational forces, and so these forces are underestimated. Include some dark matter, and it was possible to come up with a different explanation for the rotation curves, again matching exactly what is measured.

To decide between the dark matter and MoND ideas, more data was needed, and this came from measurements of gravitational lensing. Lensing is one of the unique predictions of General Relativity, which states that mass causes space and time to curve. This curvature appears as the pull of gravity, but also changes the paths of photons as they travel near a very heavy object. A region of space filled with lots of mass acts rather like a dimpled piece of glass in a bathroom window, distorting the view through to the other side. When trying to observe stars behind a heavy object like a galaxy, their image will be distorted: this is gravitational lensing. Lensing allows us to map mass in the universe without seeing that mass directly, and it provides some of the strongest evidence for dark matter in the form of the Bullet Cluster, a cluster of galaxies roughly 3.7 billion light years away.

The Bullet Cluster is really two clusters of galaxies that have collided head-on, and telescopes observe clouds of hydrogen gas bumping into each other: atoms are ionised, and X-rays are emitted. These X-rays tell us where all the regular matter is. But measurements of gravitational lensing effects tell us something else: most of the mass of these colliding galaxies has simply carried on, invisibly flying by without colliding at all. This extra mass can only be seen though these lensing effects, and when combined with the X-ray information shows that there seem to be two types of matter in galaxies: regular matter and dark matter. If dark matter is also the answer to the galaxy rotation curves, there must be a lot of it – in a galaxy like our Milky Way, there would need to be around ten times more dark matter than regular matter.

So what is dark matter? The first idea was that it is just regular matter that we can't see: some very heavy, relatively small objects that don't light up like stars – planets, black holes, asteroids, and so on. They would probably have to live in the edges (the halo) of galaxies to explain the rotational curves and other effects, so were named Massive Astrophysical Compact Halo Objects (MACHOs). The problem is that huge numbers of these MACHOs would be needed to cause the observed effects, many more than could possibly exist given everything we know about how galaxies form. MACHOs are not the answer. The top candidate for dark matter today is the opposite: WIMPs, or Weakly Interacting Massive Particles. Dark matter may be a completely new kind of particle, and the first clue to what lies beyond the Standard Model.

If WIMPs are the right answer, then we already know a few things about them: they feel gravity, but not the electromagnetic force (so transparent matter might be a more accurate name than dark matter). We know they also do not feel the strong force, otherwise they would interact with atomic nuclei. They may feel the weak force – and this would make them quite a lot like neutrinos. In fact, since we recently discovered that they do have mass after all, neutrinos actually are weakly interacting massive particles. And there are countless trillions of them flying through every galaxy in the universe, so maybe we don't need an exotic explanation for dark matter after all – it might just be the ghostly neutrino!

Of course, the answer is not so simple. Dark matter is providing the extra mass, and so the extra gravity, to hold galaxies together as they spin around. To do this, the dark matter must be moving fairly slowly; if it moves too fast, it would just fly out of the galaxy like a rocket heading out into space. And one thing we know about neutrinos is that they move fast – they are so incredibly light that they travel at almost the speed of light. Anything travelling this fast can escape galaxies with no problem, and this means that neutrinos cannot be dark matter: just like boiling water escaping from a kettle, they simply wouldn't stay in one place long enough to hold galaxies together. Dark matter can't be 'hot'.

So dark matter WIMPs must be 'cold': they must be moving slowly, and that means they must be heavy. Not too heavy, otherwise there would be another problem: it would move so slowly and clump together so well under the force of gravity that galaxies would look much smaller and denser than what we observe. Precise models of the universe suggest that dark matter is a WIMP with a mass around the electroweak scale (around 100 GeV). Nothing like this exists in the Standard Model. Dark matter is all around us, spread through all galaxies and the entire universe – pick up your favourite coffee cup, there are probably around five WIMPs drifting through it at any time – and yet we have no idea what it is.

The models give us some hope of learning more though: in order for this WIMP to have been produced in the right kinds of quantities after the Big Bang, it would probably have to feel the weak force – and this means we have a chance to detect it, just as we did for neutrinos. This is known as the 'WIMP miracle', because it suggests that experiments over the next few years should be able to measure it in the laboratory and finally answer the dark matter puzzle – and I'll say more about those experiments in the next chapter.

Dark energy

While gravitational effects give us the strongest clues to the existence of dark matter, there are also signs that gravity itself is doing something unexpected.

I find it amazing that as recently as 1925 the consensus was that our galaxy, the Milky Way, was the only one in the entire universe. Only in that year did the American astronomer Edwin Hubble make measurements of clusters of stars and nebulae visible in the night sky and prove that they must be entirely separate galaxies, much further away than at first thought. Hubble then went on to study how these distant galaxies were moving relative to us, by measuring their red-shift. This is similar to the Doppler effect: the sound of a police siren is higher in pitch when the police car is moving towards you, then lower in pitch when moving away. The light from stars appears at a higher frequency (blue-shifted) when stars are moving towards us, and a lower frequency (red-shifted) when moving away. By measuring the shifts of distant galaxies, Hubble found they were all moving away from us. If they are moving away, then at some point in the past they must have been a lot closer. And around 13.8 billion years ago, everything must have been concentrated in one place: the Big Bang, the birth of the universe as we know it. Not just matter, but space and time, expanded out from this point.

If the Big Bang was like a huge outwards explosion, then gravity is the opposite: it is always attractive, and always pulls matter together. One of the big questions in cosmology was whether gravity will eventually win out, pulling the universe back together in a 'Big Crunch', or whether the initial explosion of the Big Bang was so powerful that the universe would continue expanding for ever. But by the 1990s, this question became almost irrelevant when the red-shifts of distant galaxies revealed something completely unexpected: the expansion of the universe is speeding up. Some force must be acting to overcome gravity, pushing things further and further apart. This force is known as 'dark energy', which is really just a label for something that we don't yet understand.

Einstein had already found a way to make gravity do some unexpected things when playing around with General Relativity in 1917. This is before Hubble discovered other galaxies, and long before the idea of a Big Bang: the consensus at the time was that the universe was pretty constant – that it looks the same today as it did billions of years ago. But now there was a problem: gravity tends to clump things together, and this clumping means the universe should be changing appearance as everything moved closer and closer. So Einstein added an extra term to the equations of General Relativity, something he called the Cosmological Constant, Λ, which would push back against the clumping effect of gravity and keep the universe looking the same throughout time.

With Hubble's discovery of an expanding universe, Einstein abandoned the Cosmological Constant (he later called it his biggest blunder). But today, dark energy seems to be doing the same thing that Einstein imagined back in 1917: it is a mysterious force, pushing back against gravity, now speeding up the expansion of the universe. So the Cosmological Constant might be the right idea after all – and may not even be constant: the accelerated expansion in our universe only took off around 7 billion years ago. It is now a key part of the Standard Model of Cosmology, which is made up of General Relativity, dark energy (Λ), cold dark matter (CDM) and regular matter (which means stars, planets and us), all known as ΛCDM. This describes how the universe evolved from the Big Bang to today – and dark energy is by far the most important thing in it. The most precise surveys of stars and galaxies tell us that the universe is made up of 68% dark energy, 27% dark matter, and just 5% of regular matter – the stuff of the Standard Model of particle physics.

Particle physics might be able to say something about dark energy though. The Higgs field contains energy and is spread throughout the entire universe. The uncertainty principle says that quantum loops, pairs of particles and antiparticles, can pop in and out of existence even in empty space. In other words, empty space is not really empty on the quantum level, and it is possible that this energy is what is pushing back against gravity – the effect we call dark energy. The quantum prediction for dark energy is a little off though: a factor of 10^{120}! This is known as the vacuum catastrophe, or the worst prediction in the history of physics. If you asked me how many people there are in the world and I answered '7', I'd be off by a factor of a billion, 10^9. If you asked me how many stars there are in the entire universe and I answered '1', I'd be off by a factor of a hundred billion billion billion, or 10^{29}. But this is not even close to 10^{120}, an answer that is truly spectacularly wrong. So wrong it is probably telling us that we are a long way from understanding how the theories of quantum mechanics and General Relativity may really fit together, which means that dark energy seems likely to remain a mystery for some time. And this means the fate of the universe is also a mystery – it may continue to expand at an accelerating rate, with galaxies and stars becoming increasingly isolated. Or dark energy may 'turn off' again, allowing the attraction of gravity to pull everything back together. Until we really understand the nature of dark energy, there is no way to know.

Why are particles so light, or: why is gravity so weak?

And this brings us to the elephant in the room: gravity. It is the force of nature that is most obvious to us in our day-to-day lives, but that is only because we are sitting on this huge lump of mass we call the Earth. To individual particles, gravity is puny, and the Standard Model ignores it altogether.

The Standard Model deals with three forces, which have a range of strengths: the strong force is typically around 100 times more powerful than the electromagnetic, with the weak force sitting somewhere in between. Gravity is a lot weaker than anything in the Standard Model. It is weaker not by a factor of 100, or 1,000, but a factor of 10^{37} – 1 followed by 37 zeros. So weak that even if the LHC had been producing billions of proton collisions every second of every day since the Big Bang, it would still never have seen a gravitational interaction. Gravity is not included in the Standard Model because we simply cannot study it using particles. This is also known as the hierarchy problem, or the mystery of why gravity is so much weaker than the other forces.

Another way to phrase this question is to ask why particles are so light. If all particles carried more mass, then gravitational interactions between them could be as common as electromagnetic interactions. The point at which quantum effects and gravity are roughly equal is called the Planck Scale, and it is huge: around ten billion billion times heavier than a proton (this is roughly 20 micrograms, or about as heavy as a grain of sand – not a lot to us, but huge for a particle). Why are particles so very much lighter than this? This is particularly a problem for the Higgs boson. Because of the unique role this boson plays in generating masses, any loops that appear in Feynman Diagrams for the Higgs tend to increase its mass. Even if we accept that all the Standard Model particles are very light, the Higgs boson should still be extremely heavy, up near the Planck Scale due to these loops. But it isn't.

There actually seem to be three scales that we know of: the neutrinos, which have masses of a fraction of an eV; the rest of the Standard Model particles, from the electron up to the heaviest particle, the top quark, which have masses in the range 0.5 MeV–173 GeV; then the Planck Scale, which is around 10^{19} GeV. We don't know why there is this structure in nature, and the gap between the Standard Model particles and the Planck Scale may be full of exciting new things, or just a huge empty Energy Desert. The LHC and other experiments are making some inroads into this desert, reaching energy scales up to several TeV, but there is still a long way to go. The question is, are there any oases of new particles along the way, or does the next discovery lie far far in the distance? This is the kind of question that keeps physicists up at night.

Is there a Theory of Everything?

An even bigger question is what happens at the Planck Scale. The universe must have started out at this energy – and probably even higher – and at this point, quantum mechanics and gravity must have been playing an equal role. But we have no idea what this would look like, how to take our current theory of gravity (General Relativity) and quantum mechanics and combine them. Doing so has been the dream of theoretical physicists for decades, and may solve the hierarchy problem, reveal the true nature of dark energy, and give us a real Theory of Everything. Or at least of everything that we know about so far.

We learned more about gravity early in 2016, when the upgraded LIGO (Laser Interferometer Gravitational-Wave Observatory), which is actually the combination of two detectors in Louisiana and Washington, USA, announced the first observation of gravitational waves, one of the predictions of Einstein's General Relativity, published almost exactly 100 years earlier. Gravitational waves are ripples in space and time, so when a wave passes by, space and time get distorted: the distance between objects expands and contracts. The gravitational waves that were detected by LIGO were the result of two black holes merging into one, and this particular merger happened 1.3 billion light years away, therefore 1.3 billion years ago, and it took the gravitational waves – travelling at the speed of light – that long to reach us. The numbers involved are staggering: one black hole was 29 times heavier than our Sun, the other 36 times heavier. The gravitational waves measured by LIGO were produced in the fraction of a second before they merged, and the energy released in that fraction of a second was larger than the energy produced by the rest of the visible universe combined. But by the time the gravitational waves reached us, they caused a ripple in space just around one thousandth of the size of a proton. Detecting ripples this small is really an amazing achievement, and LIGO does it using a laser beam. The beam is split into two, sent in different directions over a 4-km detector, then recombined. Any change in the length of the detector due to a passing gravitational wave shows up in the recombined laser beam in a similar way to how the harmonics can be heard when one musical instrument is slightly out of tune with another. Comparing the signals in the two detectors showed they both picked up the same tiny ripple – equivalent to seeing something moving around 1 metre on the other side of the Milky Way – giving the conclusive proof. After this first observation, LIGO should now be able to detect several events like this every year, opening a whole new window on the universe – there is no other way to study colliding black holes!

If General Relativity is based on the smooth curvature of space and time, quantum mechanics is based on the idea that things come in discrete quantities. If we study light closely, we can see that it is in fact made up of individual lumps: photons. A quantum theory of gravity would tell us that this is also true for gravitational waves: if we could study them closely enough, we should see that they are made up of individual lumps, the boson of quantum gravity, called the graviton. The first gravitational waves observed by LIGO are exactly as predicted by General Relativity with no need for a quantum theory, but now gravitational waves are finally accessible we are certainly a step closer to making progress.

Still, we know how quantum mechanics works, and it has been used to give a successful description of every other force, so why can't we just write down the quantum theory of gravity today? First of all, there is a conceptual difference. Quantum mechanics always takes place on the canvas of space and time, the basic fabric of the universe. Particles move and interact, always against this smooth backdrop. But General Relativity is about that canvas itself. It is about how matter and energy band space and time together, and a quantum theory of gravity would say that space and time are 'quantised', that they come in lumps. Saying that matter comes in lumps – particles – is fundamentally different from saying that space comes in lumps. This is like the difference between studying clouds against a pale blue sky, and realising that the sky itself is not a solid physical thing, just a collection of photons being scattered by atoms.

Assuming you can visualise that, it is possible to build a quantum theory of gravity along the lines of the other forces in the Standard Model. This works to a certain extent, but as soon as gravity becomes important – near the Planck Scale – it breaks down. The mathematics of this quantum gravity returns infinite answers, but now even the techniques of renormalisation don't work. The infinities do not go away. A completely different approach, a completely new toolkit for understanding the quantum world, might be needed before we can really build a theory of quantum gravity, a Theory of Everything.

Beyond the Standard Model

So this is the status of particle physics right now. The subatomic world is well understood, but with dark matter we know there is another particle out there. And with gravity there is a force that does not fit the current framework. Do we need to rewrite the rules of quantum mechanics? Or is dark energy telling us that we need to rewrite the rules of gravity? These open questions make it an exciting time, not least because we are now looking for ways to connect the subatomic world with the large-scale structure of the universe: this is the domain of Particle Astrophysics. Some of these questions may be answered fairly soon, but some, like quantum gravity, may remain open for a long time. For an experimental physicist, this is a time like no other: any new discovery at this point may really lead to the next big revolution in our understanding. The final two chapters of this book are about some of the ideas for what the next discovery might be, and where it might be found.

CHAPTER 10

THE REVOLUTION STARTS HERE

The Standard Model is complete, and works beautifully – until it doesn't. We see evidence for its failings out in the universe: the lack of antimatter, the hierarchy problem, dark matter, dark energy, and so on. In order to solve some of the problems, we need new physics, physics that we haven't seen before. Not in the cosmic-ray experiments in the first half of the twentieth century, nor the increasingly powerful particle-accelerator experiments ever since. Not even in the very sensitive neutrino experiments looking for the rarest of interactions, nor in any of the myriad other particle-physics experiments making precise measurements around the world.

The problems in the Standard Model are so diverse that it's hard to know if we are looking for many different things, or one discovery that will solve them all. There is certainly no shortage of ideas though, and in this chapter I'll describe some of the biggest theories, the main candidates for the next revolution in physics.

Super-Symmetry

Solves: structure of the Standard Model, Hierarchy problem, dark matter.

I have to start with Super-Symmetry, as it is both the most loved and the most hated theory in particle physics. To some, it would be one of the most surprising discoveries about the universe to find out that it does not exist. Others can't wait to kill it and move on.

The Standard Model is built on symmetries. Gauge symmetries underpin the forces of nature, and other symmetries like CP dictate how particles behave. Looking at how the Standard Model is constructed, there are a few more possible symmetries there. Symmetries that are possible in the mathematics, but we don't yet know if they actually exist in nature. And Super-Symmetry, as the name suggests, is the big one.

There are two types of particle in the world. The fermions – electrons, quarks, and so on – make up all matter. Then there are the bosons – photons, gluons and so on – which transmit the forces. Super-Symmetery (SuSy from now on) makes things, well, more symmetric by saying that this is only half the story. That every particle has a 'super-partner': there is a boson super-partner for every fermion, and a fermion super-partner for every boson. SuSy introduces a full copy of the Standard Model, and rather than having to think up a whole new set of names, the fermion super-partners just have an 's' (yes, for super) added to the front: the electron and the selectron; the quark and the squark. The boson super-partners pick up the '-ino' suffix (I honestly don't even know why). The W and the Wino; the Higgs and the Higgsino.

Part of the motivation for SuSy is simply that it is possible, based on what we know about the mathematics of the Standard Model, and every possible symmetry seems to be realised somewhere in nature. And part of the motivation is also aesthetic. It removes the divide into two types of particle, and makes the mathematics both simpler and more beautiful at the same time. This might seem like a strange line of reasoning, as beauty is one of those things that are hard to define, but we know it when we see it. But beauty has often been a guiding principle behind new ideas in physics: for Einstein when developing Relativity, and for Dirac when taking his equation seriously enough to predict the existence of antimatter. In other words, if SuSy is possible and it makes everything much neater, it really should exist. Otherwise it seems that nature has apparently gone out of its way to avoid it, making the equations uglier at the same time, and we would have to explain why that is.

But SuSy also has more tangible benefits, and the first is an explanation for the mass of the Higgs boson. The problem here was that the interactions between the Higgs boson and the Standard Model particles (the loops) should push the Higgs boson mass up and up, far beyond the value we actually measure. If SuSy exists, the interactions between the Higgs boson and the SuSy particles would have the opposite effect: they would pull the mass of the Higgs down. These two effects would cancel each other out, and make it quite sensible for the Higgs to have a mass close to the other particles in the Standard Model. Problem solved.

SuSy can also explain dark matter. The lightest SuSy particle would be stable, in the same way the electron is stable: there would be nothing lighter that it can decay into. If this SuSy particle does not carry electric charge and is fairly heavy, then it would be an excellent candidate for the dark matter WIMP we are looking for. Another problem solved.

SuSy also suggests that the Standard Model becomes much simpler at high energies. This is part of the dream of a Theory of Everything, to unite the different forces in the Standard Model into one. This has already happened in some places: electricity and magnetism are two sides of the same coin, and the electromagnetic and weak forces are tied together by the Higgs mechanism. If all the forces are to combine, at some point they must all be the same strength, and SuSy makes this happen. The strong force becomes weaker at higher energy (the property of asymptotic freedom I discussed in chapter 4), while the electromagnetic and weak actually become stronger. SuSy tweaks this evolution slightly, making three forces converge on the same point: at some very high energy, all the forces in the Standard Model are equally powerful. We don't know if this is actually significant, and given the extremely high energies where this happens we may never find out. But it certainly suggests that the forces are somehow all related, and many attempts to build a Theory of Everything actually require that SuSy exists.

So theoretically, SuSy is compelling. SuSy should also be appealing experimentally, as it gives us a whole spectrum of new particles to discover and study. But here is the problem: the super-partner particles should have the same mass as their Standard Model partners, and if this were the case, we would have found them all already. But despite decades of searching, none have been discovered, and we actually have absolutely no direct evidence that SuSy exists. Game over. If a theory makes a prediction and that prediction is proved false, then the theory should end up in the bin.

But this is not the end of the story for SuSy. We have already met an example of a symmetry that doesn't work as expected: the W and Z bosons are much heavier than the photon, because the Higgs mechanism breaks the weak-force symmetry. Something similar might have happened to SuSy: the symmetry may be broken, making all the super-particles much heavier than their Standard Model partners and, so far, out of our reach.

This makes things more complicated, but there is freedom within SuSy for this to happen. And there are plenty of versions of SuSy that achieve it in different ways: the Minimal Super-Symmetric Standard Model (MSSM); the constrained Minimal Super-Symmetric Standard Model (cMSSM); and the Next-to-Minimal Super-Symmetric Standard Model (NMSSM). Then there is SuSy with four Higgs bosons, split-SuSy, R-parity violating SuSy, gauge-mediated, gravity-mediated or anomaly-mediated SuSy breaking. And so on. SuSy has been around a long time, and people have spent a lot of time exploring all the possibilities.

In fact, there are so many possibilities that SuSy actually has the freedom to do almost anything. The Standard Model has 26 free parameters, 26 numbers that we cannot predict, we simply have to go and measure. Some versions of SuSy have over 100. Over 100 numbers that determine how the SuSy particles will behave. So yes, SuSy predicts lots of extra particles, but which, if any, are likely to show up in experiments? What would their properties be? Could SuSy really explain dark matter? Well, it depends on where all of these dials are set. It's like saying that anyone could win the World Cup, it just depends on how all the teams play – it's true, but it's just not helpful. The flip side of SuSy's freedom to predict almost anything is that it ends up predicting almost nothing.

And this is really why SuSy has lived for so long: whenever an experiment finds no sign of the super-particles, it is possible merely to adjust some of these free parameters so that these super-particles must be just a little bit heavier, just a little bit further out of reach. By never being specific, it is never wrong. This game of cat and mouse has been going on for over forty years.

But the tide may finally be turning. Since the LHC started colliding protons at 13 TeV in 2015, we can now search for extremely heavy particles. If we still don't uncover SuSy, then it can of course still exist, but we effectively end up with a new hierarchy problem: we would need another new theory to explain why the SuSy particles are so very much heavier than the Standard Model particles. So perhaps it will finally be time to give up on SuSy in the face of an overwhelming lack of evidence. SuSy may end up as another beautiful theory destroyed by an ugly fact, and we should find out in the next five years.

Magnetic monopoles

Solves: structure of the Standard Model

There is a curious difference between electricity and magnetism. Particles can carry individual electric charges: an electron is negatively charged, a proton is positively charged. But magnets always come as pairs. Take something like a bar magnet: it has a north pole and a south pole, which are equivalent to the positive and negative electric charges. But cut the magnet in half and you don't end up with a separate 'north' magnet and 'south' magnet, you end up with two smaller bar magnets – still with both north and south poles. Keep cutting and you just get smaller and smaller magnets. It's impossible to get a single north or south pole, a magnetic monopole.

Magnetic monopoles are a bit like SuSy. If they did exist, it would make the equations of the Standard Model simpler and more beautiful, and many candidates for a Theory of

Everything include them. So perhaps we really need to think of a reason why we haven't found them yet, and the best guess is that they are extremely heavy, far too heavy to be produced at the LHC. There may be some left over in the universe from the Big Bang, but so far they have proved elusive – if they exist at all.

New forces, new particles

Solves: structure of the Standard Model, strong CP problem, dark matter

SuSy and monopoles explore some of the possibilities within the existing Standard Model. It is of course possible to add many new things to it as well.

One of the obvious ones would be a new force of nature. The forces in the Standard Model are built on gauge symmetries, with a one-dimensional charge (which is similar to electromagnetism), a two-dimensional charge (the weak force) and a three-dimensional charge (the strong force). There may be more forces: why not add a four-dimensional one, or five-, or . . .? Anything is possible, but there must also be a reason why we haven't noticed this force in any experiment yet – though the strong and weak forces provide the models for that.

A new force may be confined like the strong force, and one such candidate is known as 'technicolour'. It would show up in high-energy collisions, as some of the particles we think of as fundamental will start to break apart, just as the proton was smashed apart into quarks in 1968. Alternatively, the new force may have an extremely heavy boson like the weak force, in which case it would show up in some rare and heavily suppressed interactions, as the W first showed up in beta decay. Finally, the boson of the new force may simply not interact very often: one idea for solving the strong CP-problem involves a new force that comes with a boson (known as the axion) which is light but incredibly hard to detect – and is another candidate for the dark-matter WIMP.

There is also the possibility that dark matter might be more than one particle. There could be a whole 'dark sector' out there, something with a rich and varied set of particles, almost like a parallel copy of the Standard Model. But none of these particles carry the Standard Model charges (electric, colour or weak), and none of the Standard Model particles carry the 'dark charges' – so the two sides simply don't talk to each other. It's a strange idea – there could be a parallel world all around us, and we simply can't see it! And if this parallel dark sector really doesn't interact with the Standard Model particles at all, it would be incredibly hard to detect.

Some models unify the known forces into a larger 5- or 10-dimensional group. The forces that we currently know are just different sides of this larger group, and like archaeologists we have simply scratched the surface of what lies beneath. If we could dig deep enough, or reach high enough energies, we would see that all of the forces are really connected. Some of the larger models that unify the known forces feature a boson that looks like a heavier Z boson (the Z′) or heavier W boson (the W′). Some predict another generation of fermions, or particles like leptoquarks that are a cross between the particle types we already know. All of this is possible, but does any of it exist? Well again, we have no direct evidence.

Extra dimensions

Solves: hierarchy problem, Grand Unification

So now we have looked at ideas to take the Standard Model and make it more complete (like SuSy), or to use it as a template for new things (like extra gauge symmetries), but there are also ideas for completely changing the rules. This is usually done to try to deal with the elephant in the room: gravity.

The first example of this provides a possible explanation of the hierarchy problem, or why gravity is so weak: extra dimensions. We live in a four-dimensional world. There are three dimensions in space, each corresponding to a unique direction: up–down, left–right, forwards–backwards. There is also a fourth dimension, time, though we don't seem to have much control over how we move in time. All of the forces and particles we know move around in these four dimensions, but what if there are more? What if gravity appears so weak to us in our four-dimensional world, because most of it is leaking out into a fifth dimension, and a sixth, or even more?

First we have to solve the same old problem: if this is real, why can't we see these extra dimensions? If they do exist, something must have happened to them. One possibility is that they are curled up ('compactified') much too small for us to notice. Trying to imagine what an extra dimension would be like is a mind-bender at the best of times, without even thinking about a curled one. I tend to think about them in terms of coastlines. When seen on a satellite image or a map, a stretch of the coast tends to look quite simple: long, flat, fairly straight stretches of beach, broken up by the occasional river or inlet. Zoom in and things get more complicated: there is more detail on the beaches, smaller streams become visible, and so on. Go and stand on the beach, and you see a whole other layer of structure. The water swirls in and out around rocks, and up and down over different obstacles. This level of detail is invisible when looking at a flat satellite image, but when out walking it makes a huge difference whether you are strolling along a smooth beach or scrambling over rocky cliffs. Extra dimensions might be a bit like the rocks and pebbles along the coast: much too small to be important in the universe as we see it, which is like a satellite image. But the reason gravity seems so much weaker than the other forces could be that it has to scramble over all the rocks and pebbles, all of these curled-up extra dimensions, losing most of its power as it does so.

An alternative to compactification is brane worlds. Imagine our entire four-dimensional universe making up one floor of a building. There may be other branes, other universes like ours which make up other floors in the building, and now comes the trick: the Standard Model is working on one floor, but gravity can freely take the stairs up and down, roaming the building. I work in a building with several floors, and I see much more of the person who sits in the office next to me than of the security guard who patrols the whole building. Suppose that gravity appears weaker because it covers the whole building, where the other forces are limited to one floor. Then, as with compactification, most of the force of gravity is disappearing into these extra dimensions.

There are many versions of extra-dimension theory, dating back as far as the 1920s with the Kaluza–Klein theory attempting to unify electromagnetism and gravity into a five-dimensional universe. Renewed interest in these ideas late in the twentieth century led to the Randall–Sundrum model, the ADD model, gauge-mediated models, large extra dimensions, warped extra dimensions and others. All of these models have different strengths and solve the problem of gravity in a different way, but all use extra dimensions to do it. They would also show up in different ways at the LHC: a new heavy boson (or a whole series of them), or energy 'disappearing' into the extra dimensions, or some different particle interactions.

If it is possible to make a quantum theory of gravity, with or without extra dimensions, then there might also be quantum black holes. Black holes usually form when a massive star collapses at the end of its life. As more and more of the star's mass accumulates in a smaller area, the force of gravity increases until eventually it becomes impossible to escape, even for the photons travelling at the speed of light. A quantum black hole would be a subatomic version of this, one that would disappear almost immediately, evaporating in a shower of new particles. This would make for a rather spectacular and unique signature at the LHC, firing out many W bosons, for example.

However, this is not why most people found the idea of making black holes in Switzerland interesting. In 2008, legal action to prevent the switch-on of the LHC was brought by some concerned citizens, worried that the LHC would make a black hole and destroy the world. This black-hole story is probably the thing that comes up most often when I tell people I work at CERN and, like the name 'the God Particle', perhaps there is no such thing as bad publicity. There was a serious scientific review into the possibility of destroying the world, but in the end if it were possible to make a world-destroying black hole in a high-energy particle collision, it would have happened a long time ago. The LHC does make the highest-energy collisions in a laboratory, but outside the lab it is nothing special – much higher-energy collisions occur frequently when cosmic rays hit the Earth's atmosphere, for example. The legal process concluded that the benefits to science far outweighed the possible risks, and if you are reading this, the LHC has not yet destroyed the world.

String theory, and particle physics without particles

Solves: Grand unification, dark energy, or pretty much anything you like

There are a few theories based on getting rid of the usual idea of a particle. These vary from strange things like unparticles, which appear the same regardless of how closely we look; and chameleons, which change their properties depending on what else is nearby. But the most famous theory of this kind replaces particles with strings. String theory has been around since the 1970s, and at one time seemed to hold huge promise. Fairly or not, it is now blamed for draining years of research time and money without producing anything useful.

The idea of string theory is simple. Particles, rather than being these strange things that are infinitely small points, would have a definite size – either like a short piece of string or a tiny loop like an elastic band, around 10^{-35} m long, or a billionth of a billionth of the size of a proton. String theory has many positive aspects: it naturally describes gravity and unifies all the forces in the Standard Model. The infinities that plagued quantum mechanics before the development of renormalisation techniques no longer appear – partly because we are no longer running things all the way down to the scale of an infinitely small dot. Unfortunately, string theory also requires some fiendishly difficult mathematics – and there are a few different versions of it. A big breakthrough came in 1995 when it was shown that these were just different ways to express the same idea, and that same idea is now known as M-theory. But it's still not clear what M-theory actually is.

String theory generally requires Super-Symmetry and extra dimensions (11 in total), so if we don't discover these at the LHC then it does make accepting string theory more difficult. But the advantage and the problem with SuSy is that it could actually exist at almost any energy, so even if we don't find it at the LHC it certainly doesn't kill string theory.

In fact, there is so little connection between string theory and anything we can test in experiments that I tend to think of it as a branch of mathematics rather than of physics. Some of the mathematics behind string theory has proved useful in other areas – describing quark–gluon plasma is one – but this is not the same as string theory itself making any predictions. And this is also the reason for the current backlash: after forty-five years of work it remains very far from being a testable theory. On the one hand this is perhaps not surprising, as string theory only really becomes relevant at the Planck Scale, which is far from anything we can access in experiments, or possibly anywhere in the universe. On the other hand, I think we need much more information, more experimental discoveries closer to the energies we can reach, if we are really going to have any clue about what might happen up at the Planck Scale, and if string theory is the right answer.

The best of all possible worlds?

Solves: the unbelievable unlikeliness of the universe

This chapter has been all about the search for a deeper explanation. A theory that can answer some of the open questions in particle physics, and perhaps even a Grand Unified Theory of Everything, the ultimate instruction manual for the universe. But the existence of such a theory raises some interesting questions, because when we look around at the universe, it all seems rather unlikely that we should be here.

The problem lies in the free parameters, the 26 numbers we cannot yet explain. For example, if the W boson had been a little bit lighter, the Sun would have burned out long before we could have evolved. If the electromagnetic force were a bit stronger, fusion in stars would not have produced carbon – a disaster for us carbon-based life forms. If the Higgs mechanism did not exist at all, particles would be massless and zipping through the universe at the speed of light – atoms would not have formed, as gravity would have been unable to pull things together to form stars. However you look at it, the parameters of the Standard Model seem to be finely balanced. Move almost any of them up or down even a small amount, and it would have a dramatic effect – and almost always would result in the universe being cold, dark and empty. If the free parameters of the Standard Model are controlled by a deeper theory, then it would seem that there is a fundamental reason why the universe appears the way it does. And trying to make sense of this is one of the most mind-blowing activities in physics.

Clearing up the cause and effect is important here, because this is not an argument for something like intelligent design. Grass does not exist in order for cows to eat it; cows evolved to eat grass because it is available. The fundamental constants were not set in order that atoms could form, stars burn, and we could evolve over 13 billion years after the Big Bang on one of countless billions of planets in the universe; it just happens that the fundamental constants produced a universe where all of this was possible. And if the fundamental constants are completely determined by a deeper theory, we should probably just consider ourselves both somewhat inevitable and very lucky.

Alternatively, the free parameters may not be explained by a deeper theory, they may just be random. In other words there is nothing inevitable about the universe, we are just accidents. This is a possibility, but it is so unlikely that it is hard to take seriously, and I'll use the mass of the Higgs boson as an example. In the previous chapter, I mentioned that some quantum loops should push the mass of the Higgs boson very high. Well, it could be that all the big numbers floating around in these loops just happen to balance out, giving the Higgs boson a mass close to all the other particles. To give an idea of how unlikely this is, let's look at the stock price for the Ford motor company going all the way back to when it was first incorporated in 1903. There have been huge changes in that time: two world wars, the Great

Depression, the 1973 oil crisis, the SUV boom in the early twenty-first century and the crash in 2008. Now randomly pick two days from different points anywhere in that history, and look at the stock price at noon on those different days. Sure, it might be exactly the same, down to a fraction of a cent, but that certainly is unlikely. The chances that the numbers in the Higgs loops just happen to randomly balance out is even less likely than this.

This leads to the idea of naturalness. If a theory has to rely on excessive 'fine-tuning', a series of such extremely unlikely events in order to produce the universe we see around us, then it is extremely unlikely that this theory is correct. To go back to the example of SuSy, it could still technically exist even if the LHC does not find it, but that starts to look very 'unnatural'. Many parameters would have to be set to very specific values in order to make the SuSy particles very heavy – and this increasing reliance on fine-tuning is why many people may give up on it.

So either the free parameters in the Standard Model are determined by a deeper theory, or they are completely random. Neither of these scenarios is particularly satisfying, because the universe we live in just seems so unlikely – so fine-tuned. But there is an alternative. The fundamental parameters may be random, but not fine-tuned at all. We might live in a multiverse.

The multiverse is, well, many universes – perhaps even an infinite number. The Big Bang that started our universe may be just one of countless others, and in each of these separate universes, the fundamental parameters could be set differently. With enough universes out there, it would be inevitable that there is one that looks like ours. A slightly less mind-bending version of this idea is to assume there is only one universe, but it is much bigger than we realise. The fundamental parameters may vary from place to place throughout this huge universe, and inevitably there is one patch where they have our values.

In both the multiverse and 'patchwork universe' ideas, we can apply something called the anthropic principle. It proposes that there is nothing special about our Big Bang or our patch of universe, but also that it's no surprise that we evolved in the place where the fundamental parameters have the values they do – if they were any different, we couldn't have evolved here to enjoy them. Even though the chances of winning the lottery are tiny, when enough people buy tickets it's almost inevitable that someone somewhere will hit the jackpot – and here we are in the winning universe.

Place your bets . . .

All of the theories described in this chapter have strengths and weaknesses, but we currently have no way to tell which one – if any – is right. A huge amount of time has also been spent searching for their predictions at experiments, so far with no success. Scientifically of course this means we need to keep looking, and the data will tell us the right answer. But in the meantime, you probably want to know what I think.

The first question to ask is which of the theories are actually testable, and the answer is pretty clear. SuSy, extra dimensions and some new particles or forces of nature may all turn up at the LHC. Monopoles could be far out of reach, and string theory is so far out of reach that it is hard to imagine any way to test it.

Then which would I like? I find it hard to believe that we are really closing in on a Theory of Everything. The gap between the things we can currently test – the 'electroweak scale' around 1 TeV – and the Planck Scale is so huge, I would like to think that there are many things to be discovered between here and there. For that reason, I would be really disappointed if SuSy turns out to be the next discovery. While I can see the appeal, it almost seems too neat. And it promises to solve so many problems that we may really be left with nothing else to do – pretty much all we need is SuSy, then string theory.

However, if SuSy can really be disproved (which disappointingly also seems to be just about impossible), this would really require quite a drastic rethink in many ideas for Grand Unification. This is exactly the kind of surprise that nature has sprung on us in the past, and one of the many surprises that I hope the future has in store.

Of the other candidates, I do find the idea of extra dimensions appealing. This requires 'changing the rules' and is both more radical and would make the universe a deeper and more mysterious place – exactly the kind of game-changing idea that I hope lies in wait. As for string theory and the multiverse, well, as an experimentalist it is difficult to know what to say. We may one day know if these are real, but I doubt it will be any time soon. The experimental case for string theory will have to be indirect – just as Pauli predicted the existence of the neutrino. The only way to balance the energy in beta decay was to introduce an invisible particle, and perhaps string theory will ultimately be the only way to explain things like the fine-tuning in the universe and quantum gravity. But right now I think we are far from the point where string theory is the only possible answer.

And this brings us back to how I started this book: what we need now is new experimental data. A new discovery at the LHC or elsewhere. Something to point us in the right direction, because we are heading into uncharted territory. The coming years really are make-or-break time, and in the next chapter I'll talk about the experiments that we are pushing to the limit in the search for something new.

CHAPTER 11

THE SEARCH FOR BEAUTY

Science is not a dry collection of facts, a list of things that we know about the world. Science is an active process. It is about finding new information. Pushing into the unknown, building a new picture, a new theory about what can be happening, and then testing that theory. Science is about asking questions and searching for answers. And the trick to all great science is finding the right question to ask at the right time.

We know there are many big problems in particle physics, problems that the Standard Model cannot answer. We have many new theories that try to solve those problems, and we now need to test them: to collect new experimental data and see if any of the theories measure up. But we have to find the right question. It can't be too general – for example: Is dark matter a Super-Symmetric particle? We need to unpack this, to find a more testable question: if dark matter is a SuSy particle, then SuSy must exist. If SuSy exists, there must be many SuSy particles – one for each of the Standard Model particles. If all of these SuSy particles exist, then we should be able to find them. So instead of asking if dark matter is a SuSy particle, we first need to ask if we can find any evidence for SuSy at all.

And the same goes for every other theory: extra dimensions, technicolour, quantum gravity, and so on; there are many theories to test, and we want to test them all. But on the experimental side, we can be pragmatic. Whatever the new physics ends up being, it will come with new particles, or new interactions. The first step in testing any and all of these theories is to find a new particle, or find some of the particles we already know doing unusual things. And this is what we are doing now at the LHC, and at many smaller experiments around the world. This chapter is a guide to some of the science happening right now, to the ongoing search for the hidden beauty in the universe, and for the new physics that could explain it.

Bump hunting 2

The most obvious sign of new physics is a new particle. And the best way to find new particles is to look somewhere we haven't looked before: to go to higher energy. The LHC began its Run 2 in 2015, now colliding protons at 13 TeV. This is a huge increase over the Run 1 record of 8 TeV, and by far the highest energy ever achieved in a laboratory situation. So the hunt for new physics is on.

In chapter 7 I described the basics of how these searches work: bump hunting. Any new particle will probably decay almost instantly, and must be reconstructed from the things it decays into. This means we would start to see collisions with certain characteristics – for example two photons that add up to the mass of the new particle. Find a cluster of these collisions – a bump – and you find a new particle.

With the Higgs search, the theory did tell us roughly what to look for, and this helped narrow the search. Now we want to test many many theories, so ATLAS and CMS are each carrying out well over 100 different searches looking for bumps in many different kinds of collision: two photons, two electrons, two muons, many electrons, muons and jets, jets and missing energy (the sign of a neutrino) – anything you can think of, casting the net as widely as possible to catch anything that may be out there.

Each of these searches also has to deal with different backgrounds, different processes that can also produce two electrons, two photons, and so on, and relies heavily on theoretical predictions to describe these. There is also another complication: looking in so many places increases the chances of false alarms – something known as the 'multiple comparisons problem', though in particle physics we call it the 'look elsewhere effect'. The random nature of collisions means that sometimes the background events can just happen to come in a cluster, producing something that looks like a bump. Just as a coin can sometimes land on heads 5 times in a row – unlikely, but it can happen. But now we are not just carrying out one search, we are carrying out over 100, and this means it is quite likely that one of them will find a fake bump somewhere; flip 100 coins, and almost certainly one of them will land on heads 5 times in a row – but this doesn't mean we have discovered anything new.

There are different ways to deal with this, but the gold standard is the '5σ' requirement, when σ quantifies how much the number of background collisions may vary up or down. A bump which is five times larger than this is so big that there is only a 1 in 3.5 million chance it was caused by a random clumping of background signals. Equivalent to a coin landing on heads 22 times in a row – flip a hundred or even a thousand coins and it's still very unlikely this will happen. So find a 5σ bump, have it confirmed by another experiment, and you have a discovery.

In 2013, right at the end of Run 1, there was a search for a new heavy particle decaying to a W and a Z boson – a signature predicted by several theories. The search focused on events where each decayed to a quark and antiquark, giving 4 jets in all, and adding together the jets revealed a small cluster of events, a small bump suggesting a new particle with a mass around 2 TeV. This bump had a significance of 3.4σ – not enough to claim a discovery, but certainly interesting. And this was a real cliffhanger, as we then had a two-year break for the LHC upgrade work, and in 2015 there was a lot of excitement to see if the same bump popped up in Run 2. And, unfortunately, there is no real sign of it so far. Take the coin that landed on heads 5 times in a row and keep flipping: if there is nothing unusual about the coin it will eventually even out to roughly equal numbers of heads and tails. Small bumps come and go, and we expect to see a few more before any discovery is announced. But if a real bump shows up, well, the revolution starts here.

Hiding in plain sight

Once having found a bump, there is still lots of work to do. The next step is to figure out what is causing it, which of the many theories out there can describe it. We move from discovery to understanding.

This process is actually still going on for the bump the LHC has already found: the Higgs boson. Is it really the particle predicted by the Standard Model? Or is it a Super-Symmetric partner to the Higgs boson? Or something else entirely? In 2012 the first announcement was cautious: we had 'the discovery of a new particle, with a mass of 125 GeV'. No mention of the Higgs, because at that time we knew very little about what was really causing the bump. After six months and some more measurements, it was concluded that the new particle was 'consistent with a Higgs boson'. A further six months of data analysis produced the bolder statement that we have 'a Higgs boson, consistent with the Standard Model'.

Certainly most people expected this particle to exist, but just because the LHC delivered this present early on, we still needed to open the box and check what was inside. The Standard Model told us many things about the Higgs: it should not have any electric charge (confirmed); it should not spin (confirmed); and it should give mass to all the other particles in the Standard Model – this point is still open, and is actually one of the things that I work on.

We have measured the Higgs boson decaying to photons, to Z bosons and to W bosons. But the question is whether it also decays to fermions – in other words, does the Higgs mechanism also give mass to the matter particles, the quarks and leptons?

In the Standard Model, the fermion masses are determined by the strength of their interaction with the Higgs field: the stronger the interaction, the heavier the particles. The same is true for the Higgs boson, and means that it is most likely to decay into the heaviest particles. The top quark is too heavy – the Higgs boson, with a mass of 125 GeV, would need to 'borrow energy' via the uncertainty principle to decay to a top quark and antiquark, which weigh in at 173 GeV each. So the best chances to see the Higgs interacting with fermions are the second- and third-heaviest: the bottom quark and the tau lepton.

The Higgs boson should actually be decaying to a bottom quark and antiquark pair around 58% of the time – and yet we've never seen it happening. This is mainly an experimental problem: the LHC produces quarks and gluons constantly, so trying to find the tiny extra contribution from the Higgs boson is almost impossible. In the next few years, we should finally have enough data to do it, but it will still require all of the tricks of the data-analysis trade to make it happen. The tau-lepton analysis is a little easier, and recently the first clear sign of this decay was detected. Still, if we see the Higgs decaying to bottom quarks more often than expected – or even if we don't see it happening at all – this will tell us that the Higgs boson we discovered is something more interesting than the Standard Model prediction. Perhaps this particle is actually one of the Higgs bosons predicted by Super-Symmetry – we may have already found SuSy, and just not realised it.

Forbidden penguins

The LHC is not just about bump hunting and the Higgs boson. There is a broad range of other science to do, based on the idea of precision measurements: checking for any cracks in the Standard Model, for any hits of something unusual. Does QCD still describe the behaviour of quarks and gluons at the higher energy of Run 2? Can the gauge bosons be smashed apart into smaller particles? Checking the fine print on the particles we already know may uncover anything else that is going on.

These precision measurements can be even more powerful than a bump hunt, and the discovery of the weak force is a good example. The first clues came with beta decay, which turns a down quark into an up quark, and produces an electron – a process that was not possible by any of the known interactions at the time, and so had to involve a completely new force of nature. If we see something happening at the LHC, something forbidden by the rules of the Standard Model, again this will be a clear sign of something new. Something like penguins, for example.

This story begins in 1977. The bottom quark had recently been discovered, and the theorist John Ellis was at CERN working on a paper describing some of the more unusual ways it might decay. Like most papers on particle interactions, this one included some Feynman Diagrams illustrating a particular process – but these Feynman Diagrams looked a little different. Perhaps even a little avian. Ellis had lost a bet with the experimentalist Melissa Franklin over a game of darts, and as a forfeit he had to include the word 'penguin' in his next paper. Not easy. His solution: twist one of the Diagrams around.

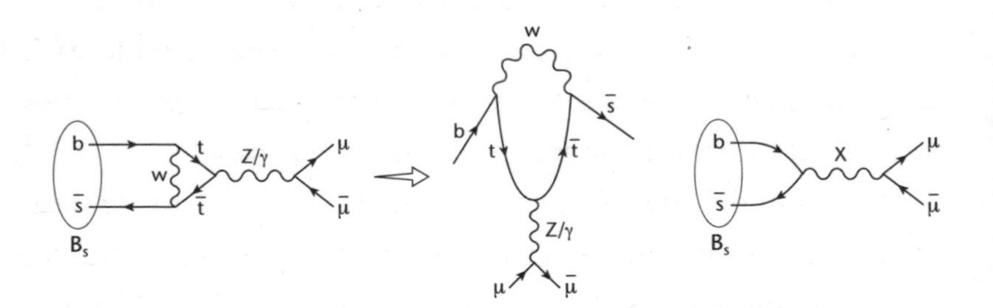

Bs→ $\mu\bar{\mu}$ in the Standard Model on the left, and redrawn to look like a penguin (sort of). On the right, an example of how this reaction might change with some new physics

This 'penguin diagram' is an example of something that is almost forbidden in the Standard Model: the B_s hadron (a bottom quark and strange antiquark) decaying to a muon and an antimuon. This process is similar to beta decay, in that it is only possible because of the uncertainty principle. The virtual particles in the diagram are much much heavier than anything else involved: beta decay typically releases a few MeV of energy, much less than the 80 GeV needed to make a W boson. So the energy must be 'borrowed', and the huge size of this loan makes the process rare. In the penguin diagram, there is a loop involving a W and top quarks, which means that even more energy must be borrowed, making this process incredibly rare. But replace the W and top quarks with some SuSy particles, or replace the whole diagram with a much simpler one involving particles from another new theory, and perhaps this process can actually happen much more frequently. Seeing B_s hadrons frequently decaying to muons would be a clear sign of new physics – and because the energy to make the new particles in the Feynman Diagram can be borrowed using the uncertainty principle, this measurement is sensitive to very heavy particles, to new physics far beyond the reach of the bump hunts.

And after decades of searching, the LHCb and CMS experiments announced in 2013 that they had observed this elusive B_s decay, but exactly at the tiny tiny rate predicted by the Standard Model, with no enhancement due to new physics. This is just one example of how sensitive the experiments have become, and how we are squeezing every piece of information out of the data, every possible clue for what lies beyond the Standard Model.

Cracks are appearing elsewhere though. The B_s measurement is just one of the many searches for rare processes described by penguin diagrams and other ways new physics can change how particles behave, and LHCb have also measured the rate at which a different kind of B hadron decays to either tau leptons or mu leptons. The Standard Model says that these two different decays are equally likely, but the measurement finds that one is more common than the other – with a significance of 3.9σ. Not quite the 5σ needed to claim a discovery, but still very interesting – and once again, the Run 2 data should tell us if this really is the first hint of new physics.

Pushing precision

Away from the LHC, there is a wide range of experiments dedicated to other measurements, testing some of the most precisely known quantities in the Standard Model. If anything unusual is going to show up, then this is a good place to look.

The ultimate rare process is proton decay. The proton is quite heavy as particles go – around 1 GeV – but in the Standard Model there is simply no mechanism, no Feynman Diagram we can draw that allows the proton to decay. Inside certain unstable atomic nuclei, it can be favourable for a proton to undergo a beta decay, converting to the slightly heavier neutron; but protons generally are completely stable. Introduce some new physics, and this can change.

Checking to see if the proton can decay essentially requires putting a lot of protons inside a detector and waiting for one very rare event: one of those protons turning into other particles. Looking for very rare events like this is something the giant neutrino detectors are very good at, and the most sensitive measurement so far has come from the Super-Kamiokande experiment in Chapter 8, which contains around 50,000 tons of pure water – a lot of protons! If protons do decay, it would be a random process following the rules of quantum mechanics: some would decay quickly, some more slowly, with the average lifetime somewhere in the middle. So if there are no decays at all in a certain period of time, then it is possible to say that the average proton lifetime must be much much longer – and because they have never seen a decay, Super-Kamiokande is able to say the average proton lifetime must be over 10^{34} years – for reference, the Universe is only 13.8×10^9 years old! The incredible stability of the proton is a good thing for all the atoms we are made out of, but has proved to be a killer for many Grand Unified Theories: when coming up with a brilliant new idea, one must be careful not to predict an interaction that allows the proton to decay.

There are many other precision measurements and searches: looking for rare muon decays, rare kaon decays, the sterile neutrinos and majorana neutrinos I mentioned back in chapter 8, and many many more. To give an idea of the diversity,

I'll mention just a handful here. The Monopole and Exotics Detector at the LHC (MoEDAL) harks back to the photographic plates used in the 1940s to study cosmic rays, but replacing those plates with sheets of plastic. The sheets are 'developed' after some time by a chemical etching technique to reveal any particle tracks burned through the plastic – the characteristic signature of magnetic monopoles being produced in collisions near the detector. The AEGIS experiment at CERN is making antiatoms to see if they feel the gravitational pull of the Earth just like regular atoms, or if they feel it as a push. There are searches for axions that involve shining a laser through a wall (really – a photon may turn into an axion, fly through the wall, then turn back into a photon), and a device called the Holometer at Fermilab which is attempting to detect quantum 'holographic noise', or the lumpiness of space and time which may result from quantum gravity.

The last one I'll mention is 'g-2', or the measurement of the magnetic moment of the muon. The muon, like all fermions, spins – and because a spinning electric charge generates a magnetic field, all fermions act like tiny bar magnets. The magnetic moment, g, is a measure of the strength of that tiny magnet; it should have a value around 2; so g-2 should be around zero. I mentioned the magnetic moment of the electrons briefly back in chapter 2, as it is the most precisely calculated quantity in QED, and QED is the most precise theory in science – and this makes it very sensitive to any new physics changing the value slightly. The muon has a larger mass than an electron, which can also enhance any possible new interactions, making this a very good place to look for a discovery.

Put muons in another magnetic field, and they will wobble a little bit – a bit like a compass needle spinning around to line up with the Earth's magnetic field. But as with the penguins or proton decay, new heavy particles can influence the interaction between the muon and the external magnetic field, and the wobble might end up being larger than expected. Measure the wobble, and it is possible to extract the value of g-2 – and possibly the signs of new physics.

This was done by an experiment at Brookhaven National Laboratory on Long Island, New York which ran from 1997 to 2001. And the muons there were wobbling more than expected: the measurement was 3.6σ away from expectation, almost as significant as the rare B-decay signal at the LHC, but still not a definite discovery. The experiment had reached the limit of its sensitivity though, and a lot more muons would be needed to improve the precision of the result. So in 2013 the experiment was moved to a place that can make them: Fermilab, Illinois. The core of the experiment is a giant circular magnet 15 m across, which was loaded onto a barge and sailed down to the Gulf of Mexico, up the Mississippi and through the Illinois canal system, then finally transferred on to a truck and driven the final stretch to Fermilab – a journey of over 5,000 km. The upgraded experiment should be online soon, so again we'll know within the next few years or so if the Brookhaven result is a real sign of new physics or not.

The search for new physics is going on in many places,

from bump hunts to penguins and precision measurements. And while there is nothing conclusive yet, there are hints, new particles or strange behaviours that could be the first sign of the coming revolution. But there is also one thing we already know must be out there waiting to be discovered: dark matter.

Hunting the dark

The search for dark matter is quite different from the search for SuSy, extra dimensions or any other theory. With SuSy, the theory came first: a neat way to extend the Standard Model, guided by mathematical beauty. With dark matter, the data came first: measurements of galaxies that can only be explained by the presence of some strange new particle. We know this stuff is out there, we just have to figure out what it is – and to do that, we need to measure it in a laboratory.

There is a strong parallel between dark matter and neutrinos. The first evidence for neutrinos was indirect – the energy being lost in beta decay. Detecting neutrinos directly is extremely challenging, but once experiments were able to do so they found a few surprises. All of the current evidence for dark matter is indirect, from galaxy rotation curves to gravitational lensing. Measuring the stuff directly is also extremely challenging, but experiments today may finally be closing in on it – and hopefully will also run into more surprises.

We will only be able to detect dark matter if it interacts with Standard Model particles – if not, it will never leave a signal in any detector we could build. The interaction might involve the weak force, or the Higgs boson (dark matter does have mass, after all), or some completely new boson; we don't have to worry about the details right now. But if such an interaction does exist then we have three ways to study dark matter, which is really just the same Feynman Diagram rotated three ways.

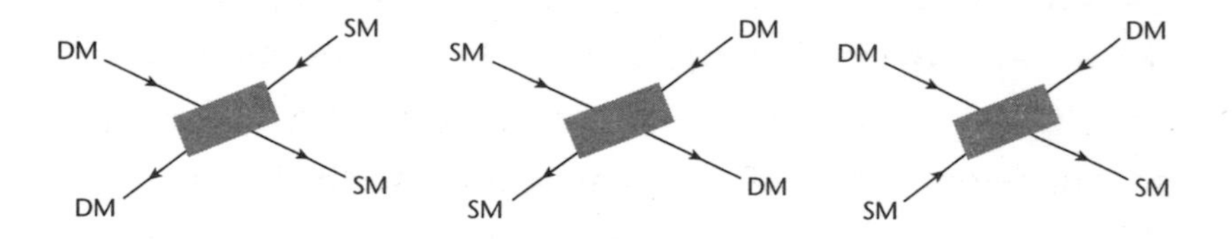

Rotating a Feynman Diagram to show three ways dark matter may interact with particles from the Standard Model. The details of the interaction are censored!

In the first, dark-matter particles collide, and turn into Standard Model particles. There is a lot of dark matter out there in the Milky Way and other galaxies, so this kind of interaction could be happening all the time. Enter the Fermi Gamma Ray space telescope. Launched in 2008, it orbits the Earth looking for high-energy photons produced in some of the most violent places in the universe: supermassive black holes, colliding stars, pulsars and supernovae. The dark-matter hunters realised it might also be picking up the signal they were looking for – and as the data from Fermi is freely available online, they took a look. Sure enough, they found many gamma rays coming from near the centre of the Milky Way – the kind of signal dark-matter collisions might produce, coming from the place where most of the dark matter lives. The centre of any galaxy is a complicated place though, with many other possible sources for these gamma rays – so this observation would have to be backed up from another angle before we can really claim to be seeing dark-matter interactions.

For the second method, if we flip this Feynman Diagram around, we now have two Standard Model particles colliding to make dark matter. And if there is one place that is colliding a lot of Standard Model particles, it is the LHC: it might be possible to make dark matter at CERN. The signal would appear in much the same way as neutrinos: dark matter would not interact with the detectors, but would leave undetected. The characteristic signal would be missing energy: an unbalanced collision with some reconstructed particles on one side and the missing dark matter on the other. The hunt for these collisions has not turned up any discovery yet, but with Run 2 under way it may come soon.

Even if the LHC does discover some characteristic collisions, it will still not be conclusive. As with neutrinos, we really want to measure dark matter directly in order to learn more about it. Rotating the Diagram again, we now have a dark-matter particle bouncing off a Standard Model particle – something that looks very similar to the neutrino interactions from chapter 8. And the 'direct-detection' experiments are based on very similar ideas to the neutrino experiments: hugely sensitive devices hidden deep underground in disused mines to shield them from background radiation and cosmic rays, looking for just a handful of dark-matter interactions in a year. The precision has been increasing, and next-generation experiments are due to start taking data in the next few years, like Xenon 1T under the Alps in Italy, and Lux-Zeplin (LZ) in the mine in South Dakota – the same mine that previously housed the Ray Davis neutrino experiment.

Ideally, a signal will appear in the direct-detection experiments and at the LHC, and match gamma-ray signals observed by the Fermi telescope. If not, then dark matter is going to be much more elusive than we had hoped.

A bigger boat?

This really is an interesting time. We have the largest, most powerful experiment in history, and many other smaller experiments pushing technologies to the limit, all in the hunt for new physics. There are already some hints, and within the next decade the LHC may have discovered a whole family of new particles, SuperNEMO may reveal the Majorana nature of the neutrino, and LZ may finally detect dark matter.

Looking a little further ahead, the LHC will be upgraded again in 2019–20, not to higher energy this time, but to a higher rate of collisions (known as luminosity). The High

Luminosity LHC (HL-LHC) may then run until 2035, and will be used to study all of our fantastic new discoveries in detail, or really flush out all the remaining corners new physics may be hiding in.

The Deep Underground Neutrino Experiment (DUNE) is currently scheduled to start taking data in 2022, carrying on the tradition of neutrino experiments in that mine in South Dakota. Studying accelerator neutrinos produced 1,200 km away at Fermilab, Illinois, DUNE should answer the remaining questions about neutrino mixing and CP violation, bringing us closer to understanding the absence of antimatter in the universe.

Looking further into the future becomes a game of speculation. If we have the first sighting of a new family of particles at the LHC, the next step may be obvious: push the energy higher and find the rest. But this is not easy; it would either require new, even more powerful magnets to keep the protons moving in the same circular tunnel, or a larger tunnel. A 100-km ring is currently being discussed, completely circling Geneva to house a machine reaching 100 TeV, almost 8 times more powerful than the LHC. Adding a new layer to the CERN palimpsest, this Future Circular Collider, or FCC as it is currently known, would be fed by protons from the LHC before accelerating them to this new record energy. China may take the lead, with plans for a similar machine that could be ready before the FCC. There are also plans for a new precision machine, colliding electrons and positrons, allowing us to study the Higgs boson or one of the new discoveries in great detail. This machine could be the first thing to go in the new 100-km tunnel at CERN, or it could be a linear accelerator somewhere else in the world.

However, it is also possible that we find nothing. The LHC may prove that, once again, the Standard Model works as perfectly at 13 TeV as it has everywhere else. All the bumps and unusual behaviours may just smooth out over time. Dark matter may remain elusive, as may the signal of neutrino-less double beta decay. DUNE may tell us everything about neutrino CP violation, but it may still not explain the absence of antimatter in our universe. And in this scenario, it is really not obvious what to do.

There would still be a case to be made for the FCC – the only way to make discoveries is still to push into unexplored territory, just as the LHC is doing today. But the LHC was really a win–win: it was guaranteed to either find the Higgs boson, or push the Standard Model to breaking point. Since finding the Higgs, we know there must be more out there, but there are very few clues about where that is – and if the searches in the coming years find nothing, there may be no guarantee of finding anything even all the way up to 100 TeV. In 1993, the US Senate cancelled the Superconducting Super-Collider, a huge machine that was under construction just south of Dallas, Texas. It would have collided protons at 40 TeV, far higher than the LHC, and found the Higgs boson many years earlier. But it was considered too expensive, and with no clues about what discoveries could be made, the plans for an even larger collider at CERN may meet the same fate. The future of particle physics may be to return to the drawing board. Instead of another huge experiment like the LHC, the future may be diverse: devising many new smaller experiments and new approaches, testing every remote corner of the Standard Model.

This is where we stand now: on the verge of a revolution, or in store for a much longer wait. This is the time for innovative and diverse experiments and data analysis, to cast the net as wide as possible, and to make sure we are asking the right questions in order to hit on the next big discovery.

The meaning of it all

The twentieth century saw a revolution in physics, which led to a revolution in technology and in the way we live our lives. Quantum mechanics came as a complete surprise, as did all of the exotic particles that followed, and it took decades to make sense of all this. The resulting theory, the Standard Model, is both scientifically and culturally important: never in our history have we had such a detailed understanding of the universe.

This book has been about that understanding – the experiments and theories behind it, and the underlying beauty in the universe that we are slowly uncovering. It has not been about all the spin-offs that happen along the way – though there have been a great number, and I have mentioned just a few throughout this book. It has also not been about how useful something like the Higgs boson is – and that is simply because I really have no idea. It is almost impossible to imagine the directions that new discoveries take us: nobody could have foreseen quantum mechanics leading to semiconductors and the digital electronics revolution, for example. In 1943, the president of IBM thought: 'There is a world market for maybe five computers.' I am certain, however, that if we don't make these discoveries, then they will never be used – and that the discoveries we have already made will be used somehow in the future. Right now I might think that there is no world market for Higgs bosons, but people will hopefully look back and find this statement hopelessly naive as they jet around in their Higgs-powered spaceships.

There also is an entirely different story to be told about the people who, like me, devote their lives to studying the infinitesimally small. There are a lot of us, and each one will have a different story, a different moment when we got hooked, and a different reason for thinking this is all worthwhile. So I'll just give you my opinion. I think that the world is a better place for knowing that the Higgs boson exists. If there are extra dimensions, then the world will be a better place because we know that too. Better in the same way that the world is a better place because the *Mona Lisa* exists, the Taj Mahal, the pyramids in Egypt, Machu Picchu, *Hamlet*, General Relativity, music, dancing, the internet, or anything else that makes our lives richer, more fulfilling, and more beautiful. I can't think of anything I would rather do than help to write the next chapter in this story. And that next chapter may be just beginning . . .

CHAPTER 12

THE NEW PHYSICS

In 2015, the LHC saw the first-ever collisions at 13 TeV. It was a slightly rougher ride than expected, but bringing this fantastically complicated machine back online after two years of repairs and upgrades, then pushing the performance to deliver collisions at almost twice the previous record energy, was never going to be easy. But from June to November, protons were being smashed and the results recorded.

There was also plenty of work for the experimenters: updating the calibrations, learning about the behaviour of the detector and the collisions themselves at these new high energies. And, of course, analysing those collisions looking for signs of something new. To avoid biasing each other, the ATLAS and CMS experiments work independently until they are confident in the results and ready to make them public. And by 15 December 2015 they were ready to compare scorecards.

To ensure the independence of the experiments, there has to be a degree of secrecy. The data each experiment records, and the status of ongoing analyses, is kept behind closed doors until the results are ready to be made public. But with over six thousand people involved, rumours do get out. And in the run-up to 15 December there was a rumour that ATLAS had found something unexpected.

Any discovery at this point is huge. The Standard Model does not predict any more particles, so if a new one shows up then it would be momentous – the news people have been waiting for years, for decades, to hear. And so some theorists got a little overexcited. We call this 'ambulance chasing': any hint of something new, no matter how small or insignificant, and someone will jump on it as conclusive proof of a pet theory. And having heard the rumours, a lot of people were ready to jump.

I work on ATLAS, so I followed some of the internal discussions before the results were public. There was a bump. Not a huge bump, but a bump nonetheless. A clustering of events that contain two photons – the same kind of events that we used in order to discover the Higgs-boson bump I described back in chapter 7. But this bump was much heavier: the Higgs boson has a mass of 125 GeV, this new bump was at 750 GeV. Upon seeing something like this, there was excitement and scepticism in equal measure. Surely this was just a random clumping of background events. Or some noise in the detector. Many cross-checks were carried out, but the bump did not go away. And so, on 15 December, it was time to go public, and carry out the ultimate cross check: did CMS see the same thing?

The seminars were held in the main CERN auditorium, the one used to announce the Higgs-boson discovery back in 2012. This time the talks were scheduled for the afternoon, so no need to camp out overnight, but the hall did start filling up a few hours in advance. And after showing many of the impressive new measurements that had been carried out with the data, there came the histogram everyone wanted to see: take the events with two photons, and combine the photons to calculate the mass. And ATLAS and CMS both had a bump. In the same place: two photons adding up to 750 GeV.

The official CERN announcement was understandably cautious, stating that the two experiments had 'observed a slight excess in the diphoton decay channel'. And taken on their own, the two results are not enough to get too excited about: the CMS bump was 2.6σ over the background, and the ATLAS bump 3.6σ. There is an explanation of σ in chapter 7, but it is a measure of uncertainty, a way to quantify how likely it is that a random clumping of background events may produce a fake bump. The higher the number, the less likely it is to happen, and we normally require 5σ over what we expect from the background before claiming that a bump must be the sign of a new particle. The fact that ATLAS and CMS see a small bump in the same place certainly makes this more exciting – this is exactly how a new discovery would start to show up – but so far it is not definitive. We would need more data to tell if it was real or not.

But this was already enough for some. Within minutes of the announcement, the theory papers started to appear – so quickly that these first papers must have been written in advance of the announcement, based just on the rumours. Eight on the first day; 65 after a week; 121 by the end of 2015. As the LHC returned from the winter break early in 2016, there were over 300. I certainly haven't read them all, but doubt that there can really be more than 300 interesting things to say about this bump already. The deluge of papers is probably due to a combination of overexcitement, a scramble for precedence, and the pressure from funding bodies to be continuously producing papers.

And to be fair, if this really is a new particle, it is very exciting. And not least because it is also fairly unexpected – of all the things that could have turned up, not many people would have bet on a new two-photon signal. It can also be interpreted in many ways: it could be a particle associated with extra dimensions. It could be a completely new boson, associated with a previously unknown force of nature. And, of course, it might be one of the particles predicted by Super-Symmetry, because that theory can be made to predict almost anything. Or it might be a second, heavier, Higgs boson – which would be a total surprise.

Both experiments were also busy over the winter break, performing a more thorough analysis of the 2015 data using the latest calibrations and analysis techniques. And the bumps got a little bit larger: 3.4σ for CMS and 3.9σ for ATLAS. Still not definitive, but definitely moving in the right direction.

But there is also the very real possibility that this is nothing at all. Not a revolutionary discovery of a new particle, just a very suggestive, random clump of background events. Bumps like this have appeared and disappeared before. The only way to know for sure if this one is here to stay is to collect more data, and so as the LHC was ramping up to deliver that data in May 2016, it was in the headlines around the world again. But this time because of a different bump: a weasel (actually a beech marten) had found a way into one of the electrical stations above ground and gnawed through a power cable, causing a short circuit that needed a couple of days to repair. Nature has many ways of surprising us!

As I'm writing this, the 2016 run of the LHC is well under way, and when it is done and the data analysed we will know for sure if this bump is real. This is the most exciting year in particle physics for a long time – and if there is a new particle there, it will be even more significant than the discovery of the Higgs boson. It would be the first real sight of what lies beyond the Standard Model. The New Physics at last.

APPENDIX 1

EXPLANATION OF NUMBER NOTATION

Particle physics contains ridiculously large numbers of ridiculously small things. Talking about all this becomes tricky – there are only so many times you want to say 'millionth of a billionth of a billionth' of something. So we have a shorthand to write this stuff – some of which you will probably already know from computer memory: megabytes, gigabytes, and so on. Each prefix (mega-, giga-) can be shortened to a single letter (M, G) and represents an extra factor of 1,000. So a megabyte is one million bytes, and a gigabyte is one billion bytes.

These numbers can also be represented as powers of ten. Ten squared, or 10 x 10 = 10^2 = 100. Ten cubed, or 10 x 10 x 10 = 10^3 = 1,000. 10^6 =1,000,000, and so on – the 'power', indicated by the superscript, is the number of zeros if writing out in full. We can also get very small numbers when dividing by very big numbers, and this is represented with a minus sign on the power. For example, $1 \div 100 = 10^{-2} = 0.01$. And: $1 \div 1{,}000 = 10^{-3} = 0.001$, and so on.

Here is the list of abbreviations used in this book:

number	word	power of 10	prefix	symbol
		10^{15}	Peta	P
	trillion	10^{12}	Tera	T
1,000,000,000	billion	10^{9}	Giga	G
1,000,000	million	10^{6}	Mega	M
1,000	thousand	10^{3}	kilo	k
0.001	thousandth	10^{-3}	milli	m
0.000001	millionth	10^{-6}	micro	μ
0.000000001	billionth	10^{-9}	nano	n
	trillionth	10^{-12}	pico	p
		10^{-15}	femto	f

APPENDIX 2

THE SYMMETRIES OF THE STANDARD MODEL

Chapter 6 explored some of the ideas of symmetry in particle physics, and in this short appendix there is more information on the construction of the Standard Model, and the U(1) xSU(2)xSU(3) notation sometimes used to describe it. This is a little more mathematical than the rest of the book, but ultimately the mathematics is our guide to the subatomic world. And when building the Standard Model, the maths revealed a problem.

First of all, back to the idea of symmetry: a transformation that does not change the way the universe behaves. Some of these symmetries can appear to be almost trivial: a 100m running track is 100m long even though the Earth is rotating – and taking the whole track with it. To phrase this in the language of symmetries, the running track is symmetric (unchanged) under a transformation (the Earth rotating). It's possible to form a full 'group' of transformations for this symmetry: the Earth rotating, the Earth orbiting the Sun, the Sun moving around in the spiral of the Milky Way, and so on. This kind of symmetry makes a race much simpler: you just have to run from the start to the finish, without working out your orientation relative to the Sun, or the rest of the universe. When there is no symmetry, things become much more complicated: the Apollo missions to the Moon did have to account for the orientation of the Earth relative to the Moon, and how that would change during the rocket's flight.

Symmetries had been studied mathematically before being applied to physics. The nineteenth-century Norwegian mathematician Sophus Lie (pronounced 'Lee') was interested in the properties of groups of transformations, and his work led to that of Emmy Nöther and many others. So when symmetries were applied to particle physics, the tools to understand them, known as Lie Algebra, were already available.

The Standard Model is built on gauge symmetries, which relate to a quantity we cannot directly measure, like the reading on a particle's 'pedometer', or the direction of a quark's colour charge 'weather vane'. As we can't measure these things directly, their exact settings cannot change the way the universe behaves – and this means there must be a symmetry here: it must be possible to change the setting of a particle's 'pedometer' without changing the result of an experiment. If experiments were sensitive to this setting, then we would be able to measure this setting. What actually matters is not the absolute setting of a pedometer, but the relative setting between two particles – so if these settings are going to change, there must be a way for particles to communicate and work out their relative synchronisation.

Turning this into the language of Lie Algebra, a particle's internal pedometer is a 'one-dimensional' property, like a clock with one hand, and it can only be wound forward or back (not 'left' or 'up', for example). This means there is only really one possible transformation here: wind the hand around. This transformation makes up a group called the U(1) group ('you-one'). U means unitary, which means that the length of the pedometer hand does not change, only the direction it is pointing; and 1 denotes that it relates to a 1-dimensional quantity. In any unitary group, the number of associated changes is given by n^2, where n is the number of dimensions – so this group contains one change: wind the clock.

But now comes the amazing part: the mathematics tells us that there is one possible transformation for a particle's pedometer, and this one possible transformation corresponds to one real physical particle. This particle is the gauge boson for this symmetry, and it is the way that particles communicate and work out their relative synchronisation. At first, this particle was thought to be the photon, but in the Standard Model this particle is actually the B boson, which later becomes mixed up with the W^0 boson to produce the photon (and the Z boson).

For the strong force, there are three possible dimensions (red, green and blue) and more possible changes: rotate from red to green, rotate from blue to red, and so on. Gathering together every possible rotation of the colour-charge weather vane gives us the SU(3) group ('ess-you-three'). U again for Unitary, meaning the length of the colour weather vane does not change, just the direction it is pointing. The 3 because we are dealing with a 3-dimensional quantity. S is Special, meaning it does not contain something equivalent to the simple winding of the U(1) group; removing this gives (n^2 - 1) possible transformations in the Special groups.

Now again, the mathematics actually corresponds to real physical things: there are eight possible transformations in the SU(3) group, and these correspond to 8 different kinds of gluon, each carrying different combinations of colour (blue+antired, red+antigreen, and so on).

Finally, the weak charge has two settings, heads or tails, so the possible transformations here form the SU(2) group. The Lie Algebra, the mathematics of symmetry, tells us that again there should be (n^2 - 1) possible transformations, which means there should be 3 W bosons: the W^+, W^0 and W^-. And this was the clue that something was missing: the W^0 had never been seen in nature, even though the W^+ and W^- had. Explaining this required the electroweak model, which uses the Higgs mechanism to mix up the W^0 with the B – giving the photon and Z – while also giving mass to particles.

So the forces of the Standard Model are often written using the Lie Algebra shorthand: U(1)xSU(2)xSU(3). The first part, U(1)xSU(2), is the electroweak model, combining a one- and a two-dimensional symmetry; then SU(3) is the three-dimensional symmetry of QCD, the strong force. There is a group of transformations that are possible under each symmetry, and each of those transformations corresponds to a real particle, the gauge bosons that transmit the forces. The gluons transform the colour state of quarks, the W bosons transform a 'heads' particle to a 'tails' one (like electron to neutrino), and the B boson winds a particle's pedometer. This is not just mathematics, these are real events. The forces in the Standard Model, the forces that shape the world around us, are really due to these gauge symmetries, which appear to be fundamental properties of the universe, all written in the language of mathematics. Beautiful, really.

GLOSSARY

atom The basic building block of all materials. Atoms consist of a cloud of electrons orbiting a tiny central nucleus made up of protons and neutrons. In a normal atom there is one electron for every proton. The number of neutrons can vary, and some configurations are unstable (radioactive).

baryon A composite particle, made up of three quarks bound together by gluons (or three antiquarks, in the case of an antibaryon).

boson A type of fundamental particle, the bosons are the 'force carriers'. The photon, the W and Z, the gluon and the Higgs are all bosons. Named for the Indian physicist Satyendra Nath Bose, all bosons spin at a rate equal to an integer unit: the Higgs has zero, and all the rest have 1 unit of spin. The theorised (and as yet undiscovered) graviton would have 2 units of spin. Bosons are characterised by the fact that, unlike fermions, they can overlap, occupying the same space at the same time.

cosmic ray Cosmic rays are high-energy particles (mainly protons) produced in some of the most violent events in the universe: exploding stars (supernovae), quasars and active galactic nuclei. When these particles hit the Earth's atmosphere they create a shower of particles, some of which (mainly muons and neutrinos) live long enough to reach the surface of the Earth. These collisions can occur at energies many thousands of times higher than those achieved at the LHC.

Dirac Equation Discovered by Paul Dirac in 1928, the Dirac Equation describes the behaviour of all fermions and is the basis for our understanding of how particles behave. Here is its simplest form: $i\gamma.\partial\psi = m\psi$

electromagnetism One of the forces in the Standard Model. Experiments in the nineteenth century proved that the apparently different phenomena of static electricity and magnetism were two effects produced by the same force: electromagnetism. In the Standard Model, the electromagnetic force is transmitted by photons flying between particles carrying electric charge. Opposite charges (positive and negative) attract, while like charges repel.

electron One of the fundamental particles in the Standard Model. Discovered in 1897, the electron carries one unit of negative electric charge, and can be found in all atoms. The interactions between electrons in neighbouring atoms determine how different chemical elements interact, and electrical power is just the movement of electrons.

electron-volt (eV) The standard unit of energy, momentum and mass used in particle physics. 1 eV is the energy gained by an electron when accelerated by an electric field of 1 volt. Energy can be translated into mass by the equation $E=mc^2$, and using 'natural units' where the speed of light is defined to be 1. The mass of a neutrino is less than 1 eV; the electron is 511 keV (511,000 eV); the proton is 938 MeV (938,000,000 eV); the Higgs boson 125 GeV (125,000,000,000 eV); and the heaviest particle, the top quark, is 173 GeV. The LHC currently collides protons at an energy of 13 TeV (13,000,000,000,000 eV).

electroweak model Developed in 1968, the electroweak model is the heart of the Standard Model. In this theory, the electromagnetic and weak forces are bound together by the Higgs mechanism, giving mass to the W and Z bosons, while leaving the photon massless.

fermion A type of fundamental particle, the fermions are the 'matter particles'. All quarks and leptons are fermions. Fermions are named after the Italian-American physicist Enrico Fermi, and obey the Dirac Equation. Fermions are characterised by the fact that they have half-integer spins (all fundamental fermions have ½ a unit of spin). Fermions also stack up: two identical fermions cannot occupy the same position (technically the same quantum state, which means having all the same quantum numbers), and so electrons in atoms do not just sit in the orbit closest to the nucleus, but fill up successively higher orbits. There are actually two kinds of fermion: all the fermions in the

Standard Model are Dirac fermions, though neutrinos may actually be Majorana fermions, which would mean they are their own antiparticle.

Feynman Diagram Cartoon that is a real representation of the underlying mathematics of particle physics. They can be used to calculate the probability of a given interaction, and all of the possible outcomes. Each Feynman Diagram represents one of the many possible things that can and do happen when particles interact.

fundamental particle A particle that is not made up of smaller particles – or at least, not as far as we can tell. Electrons are fundamental, but protons are not: they are made up of quarks. The Standard Model is the theory of the 12 fundamental fermions, along with the 5 fundamental bosons.

gauge symmetry One of the most difficult, but also most powerful, concepts in particle physics, a gauge symmetry refers to an arbitrary label used within the Standard Model. Taking the example from chapter 6, time zones are arbitrary labels, and vary from place to place and time to time. These labels are useful, but they do not change the underlying physics: how quickly the Earth rotates, for example. Within the Standard Model, the arbitrary labels are the settings of particle 'pedometers', the direction of colour-charge 'weather vanes', and the 'heads' or 'tails' of the weak-force charge. The forces in the Standard Model operate in order to preserve the symmetry: allowing these arbitrary labels to change without changing the way the universe behaves.

gluon The gluon is a boson, the 'messenger particle' of the strong force. The gluon carries energy and momentum between particles carrying the strong force 'colour' charge.

gravity The force that we are most familiar with, but that we have not yet fitted into the Standard Model. The pull of gravity keeps us here on the surface of the Earth, and keeps the Earth orbiting the Sun. The current theory of gravity is Einstein's General Relativity, which describes this force as a bending of space and time. There is currently no quantum theory of gravity, and devising this theory has been one of the goals of theoretical physics for decades.

hadron Any particle made up of quarks. There are two main kinds of hadron – baryons (three quarks) and mesons (a quark and an antiquark) – though more exotic and extremely unstable hadrons can also exist, such as tetraquarks (two quarks and two antiquarks) and pentaquarks (four quarks and an antiquark.) Over a hundred different hadrons are known, and they can be formed of any combination of quarks apart from the top quark, which is so unstable it will decay before it can be bound up.

Higgs boson The final piece of the Standard Model, discovered in 2012 at the Large Hadron Collider. The Higgs boson is the particle associated with the Higgs field, a 'store' of energy that fills the entire universe, and is analogous to a 'ripple' in that field. Other particles can interact or 'stick' to this field and pick up some of that energy as mass. The Higgs boson itself can interact with any particle that sticks to the Higgs field – ie any particle with mass.

Large Hadron Collider (LHC) The LHC is currently the largest and most powerful particle accelerator in the world. It consists of a 27-km ring buried in a tunnel just outside Geneva, Switzerland. The LHC usually collides protons at the record energy of 13 TeV, and for part of each year collides lead nuclei.

lepton Derived from the Greek for 'light' (i.e. not heavy), leptons are fundamental fermions that do not feel the strong force. Electrons, muons, taus and neutrinos are all leptons.

meson A hadron made up of a quark and an antiquark, bound together by gluons.

neutrino The most mysterious particle in the Standard Model, first predicted by Wolfgang Pauli in 1930 but not detected until 1956. There are three 'flavours' of neutrino (electron, muon and tau-neutrinos), and they have a tiny mass, over a million times lighter than the next-lightest particle, the electron. Neutrinos feel only the weak force, and therefore interact very rarely. It is not yet known whether the neutrino and the antineutrino are actually the same particle (this would make it a Majorana fermion; see chapter 8).

neutron One of the constituents of the atomic nucleus, the neutron is a baryon – a hadron made up of three quarks (up, down, down). The neutron has no electric charge.

particle The smallest piece of matter, and in particle physics there are two kinds: composite and fundamental. The composite particles generally behave like solid objects, but are actually made of smaller things. Protons and all hadrons are composite particles, being made up of quarks held together by gluons. Hit a proton hard enough, and it will break apart. Fundamental particles are, as far as we can tell, not made of anything smaller (though this might just mean we haven't hit them hard enough yet). The twelve fermions and five bosons of the Standard Model are the only fundamental particles we have discovered so far.

photon The photon is the particle of light. It is a boson, the 'messenger particle' of the electromagnetic force. The photon carries energy and momentum between charged particles, pushing them together or apart.

Planck Scale The energy and mass scale at which quantum mechanics and gravity become equally important. The Planck Scale is roughly 10^{16} TeV, or a million billion times higher than the energy reached at the Large Hadron Collider.

proton One of the constituents of the atomic nucleus, the proton is a baryon – a hadron made up of three quarks (up, up, down). The proton carries one unit of positive electric charge.

QCD See **quantum chromodynamics.**

QED See **quantum electrodynamics.**

quantum chromodynamics, QCD The gauge theory of the strong force, associated with a gauge boson: the gluon. The strong force holds quarks together in hadrons, and holds protons and neutrons together in the atomic nucleus. This force is experienced by all particles that carry colour charge (namely both quarks and gluons).

quantum electrodynamics, QED The quantum theory of the electromagnetic force. This is associated with a gauge boson, the photon, and is experienced by all particles that carry electric charge.

quantum field theory The version of quantum mechanics used in the Standard Model. Quantum field theory is built on the Dirac Equation, the Special Theory of Relativity, and gauge symmetries.

quantum mechanics Quantum mechanics describes the behaviour of the world on the smallest scales we know. The idea itself is simple: things come in small 'lumps', or 'quanta'. This applies to matter, which can be broken down into atoms and eventually into the fundamental particles of the Standard Model. But it also applies to energy and momentum, which is transmitted between particles in 'lumps', carried by bosons. The full implications of these ideas require a picture of particles that behave in many strange ways, explored in chapter 2.

quark A type of fundamental fermion that feels the strong force (as well as the weak and electromagnetic). There are 6 quarks: up, down, strange, charm, bottom, top. The name rhymes with 'bark', not with 'fork'.

relativity Einstein's theory of space and time, which comes in two versions: the Special and the General Theory. The Special Theory, published in 1905, is based on the idea that nothing can travel faster than the speed of light (c), and leads to the famous relationship between energy (E) and mass (m) $E=mc^2$. The Special Theory is valid for motion at constant velocities, and the General Theory extends these ideas to include accelerated motion – and in doing so, describes the force of gravity as a bending of space and time.

Standard Model The current best theory of the universe on the smallest scales. The Standard Model contains 12 matter particles (fermions), which interact through 3 different forces (the electromagnetic, the strong and the weak). Particles acquire mass through interacting with the Higgs field, which has another particle associated with it: the Higgs boson.

strong force, or strong nuclear force The force that holds the atomic nucleus together, and holds quarks together inside hadrons. In the Standard Model, the strong force is described by QCD, and transmitted by a gauge boson, the gluon.

Super-Symmetry Super-Symmetry (SuSy) is one of the most common theoretical extensions of the Standard Model, and predicts the existence of a 'super-partner' for each particle in the Standard Model. Fermions have a boson 'super-partner', and bosons have a fermion 'super-partner'. SuSy could solve several problems in the Standard Model, but we currently have no evidence for its existence.

uncertainty principle First introduced by Werner Heisenberg in 1927 as one of the consequences of quantum mechanics, the uncertainty principle has two forms. The first says that we cannot know both the position and the momentum of a particle perfectly; as a result, squeezing a particle into a smaller space means it tends to pick up momentum and escape. The second says that 'virtual particles' can borrow some energy for a small amount of time: the more is borrowed, the quicker it must be returned.

virtual particle A particle trapped inside a Feynman Diagram. It is impossible to measure these particles directly, and they are allowed to use the uncertainty principle to bend the laws of physics: borrowing energy for a short period of time.

weak force Probably the most mysterious force in the Standard Model, the weak force is transmitted by two bosons: the W and the Z. Unlike every other gauge boson we know, the W and the Z have mass. The electrically charged W bosons can change the type of particle they are interacting with (electron to neutrino, up quark to down quark, for example). The Z boson was predicted by the electroweak model, and behaves much like a heavy photon.

WIMP At present the favoured explanation for dark matter, a WIMP is a Weakly Interacting Massive Particle.

INDEX

Note: page numbers in **bold** refer to illustrations.

INDEX

INDEX